March 17, 2004

Expanding Perception

Consciousness, Wisdom, Compassion, Love

Klaus Heinemann, Ph.D.

For Stella Doran,
with love & blessings

Published by *Word Association Publishers*
205 Fifth Avenue, Tarentum, Pennsylvania, 15084
800-827-7903

Printed in the United States of America

Library of Congress Cataloging-in-Publication Data

Heinemann, Klaus, 1941-
Expanding Perception,
Consciousness, Wisdom, Compassion, Love
by Klaus Heinemann – 1st ed.
ISBN: 1-932205-97-7
Library of Congress Control Number: 2004101041

Dedication

To our children ... and their children ... and their children ...

Acknowledgement

The number of people who have contributed, directly or indirectly, to this book is large, indeed! First and foremost, I want to thank Adelgund (Gundi) Heinemann, an educator, alternative healing arts practitioner, and my companion and wife of four decades, for her continued inspiration, encouragement and help during the countless hours of writing and editing the manuscript. Then I wish to express my deep appreciation to my spiritual teachers, Emilia Rathbun and her late husband Dr. Harry Rathbun, inspiring wisdom teachers, for introducing *Jesus as Teacher* to me; to Dr. Ron Roth, modern-day mystic and spiritual healer, for incessant teachings about love and insights into the *spiritual reality*; and to Hermann Berning, Elisabeth Quellenberg, Magdalene Roth, and other family members and friends who have passed to the spiritual realm, but not before inspiring me with their wisdom about consciousness, compassion, and love that eventually led to the inspired writing of this book; and to many friends and relatives who have marked their contribution with their interest, advice, and wisdom.

I especially thank my friend Ben Young for many stimulating discussions in the general subject area of expansion of consciousness and valued contributions to Chapters 3.3 to 3.6.

Contents

[1] Asterisks (*) denote chapters with detailed, somewhat scientific contents that the reader who is more interested in conclusions and personal applications may wish to skip (see comment on p.10).

CHAPTER 3: Expansion of Perception

CHAPTER 4: Expansion of Consciousness

CHAPTER 5: Expansion of Belief Systems

Context

Doubt is not the opposite of faith; it is one element of faith.

Paul Tillich

"Our Father, who is in Heaven!" These first five words of the most frequently spoken prayer in the Christian tradition contain two words that trigger reactions in, I would assume, just about every person of Christian heritage. Who is the "Father"? What is meant with that word? What is "Heaven"? Where is it? Does it even exist, now that we know that there is nothing but "empty" space beyond the Earth's atmosphere? Isn't it *primitive* to still pray that way? Are we praying to a *person*? The questions go on and on. Often they trigger criticism of the Christian faith at large; often they become a stumbling block.

This book is about uncovering the deep meaning behind these words. It is about purpose. The purpose of life, of my life, of your life, the purpose of the contextual setting in which our lives can unfold: the planet and beyond.

To me, this purpose is described with one simple word: love. But humans are complicated beings. One word will not do. We need descriptions, refinements, arguments and counter-arguments. With these we develop a rationale that allows us to tailor a simple instruction for human life – to be a loving human being – into a complex societal structure that permits us to "legitimately" add one little word to the instruction, the word "self." Loving turned into self-loving. Our entire culture is built on this principle. And the more we get entrenched in it, the more we miss the real purpose.

Complexity has its positive sides. The more complex something is, the higher is its potential. Along with the enormous complexity of the human mind comes the potential for highly significant impact, effect, or outcomes it can have. These outcomes can be of cosmic magnitude.

In this book I elaborate on a personal belief system that addresses the potentiality of the human mind. In a nutshell, it is based on the following assumption:

There is, in addition to our physical world, a nonphysical reality[2]. The two are *dualism aspects* of a whole, a *holistic oneness*, which I equate with "God." The nonphysical reality is the realm of thought and consciousness.

The process of working with this assumption has expanded my perception. It has helped me gain an expanded understanding of the context and purpose of life. It has taught me about the progression of consciousness, wisdom, compassion, and love, one building on the other, ultimately leading to love as the highest of all human expressions. And it has contextualized the "Supreme-Original Entity" ("Father") and the "Spiritual Reality" ("heaven") to the extent that the old words no longer bother me.[3] They have a new meaning, a much deeper meaning than I had ever believed I would find.

---

[2] As I will elaborate in this book, both realities are actually of a "physical" nature. All depends on the definition of the word "physical." If we define the realities as realms in which processes occur in association with a *flow of energy*, then both realities are physical. If we define only that as physical which can be touched, seen, heard, smelled, conventionally calculated and measured, the reality in which consciousness reigns is not physical. For reason of simplicity of the concept, I use the latter definition in this book, except where specifically noted otherwise.

[3] I must, however, admit that my having moved to a new language environment as an adult person (from German to English) has helped in this process of assimilation of new concepts with old words. The German word "*Himmel*" (heaven), for example, still tends to give me the shivers.

Introduction

During the course of the last century, our scientific knowledge has increased to the degree that a fundamental attitudinal change has taken place among foremost scientists. A feeling of superiority, of euphoria that we were getting closer and closer to a full understanding of the intricacies of our physical world has given way to a realization that our universe is far greater, more magnificent, and bears more mystery than we had ever before imagined. With every increase in understanding how our physical universe works, a multitude of new, unanswered questions opened up. New limits were discovered. In many cases, these limits have presented themselves as limits of *discoverability*, not just limits of understanding.

The human reaction to mastery of a concept or skill is generally feeling good about oneself. When a student passes a difficult exam, he feels on top of the world. When a biologist discovers a new vaccine, he may experience something like euphoria about having made a major contribution to mankind. When we saw the first man walk on the moon, many felt as if proof had been established that God does not exist. But when we discover *non-discoverabilities*, when we find out the limits of our capacity to discover, heal, understand, then we begin to grasp the vastness of the All in comparison to the smallness of ourselves, then we become humble ... and then we have the opportunity for re-discovery of spirituality.

The discoveries in physics during the 20th century have been so revolutionary that it is justified to say that we have come full circle in the way we view metaphysics. The circle began with a realization, millennia ago, that there was so much we did not and could not understand that we had no problem believing in supernatural forces. These forces would direct human lives. Then we began the process of discovery of how things work. We became able to explain more and more of what had formerly been mysteries to us. Culminating in the late 19th and early 20th centuries, we thought we knew – or were capable of finding out – essentially everything there was to know. Concomitantly, belief in supernatural forces was seen as a sign of weakness, metaphysics was ridiculed.

Now, having come again -- albeit at a deeper level -- to the limits of our human understanding, we know that there is a *world of consciousness* beyond ("*meta*") our understanding of reality, and metaphysics has, once again, become legitimized.

In fact, one of the great realizations with which physicists appear to have entered the third millennium is the *admission* that there is more to existence than can be explained with conventional physical. To consider the physical reality, i.e., that which we can explain with the laws of conventional physics, as the entirety of existence is far too limiting and leaves increasing numbers of questions unanswerable. What *is* energy? Of course, we can describe it, we can use it, we can express it with equations, we can transform it – but what *is* it *really*? Where does it come from? What *is* gravity? We may be able to explain why it is proportional to the masses and inversely proportional to the square of the distance between two bodies, but *why is it* to begin with? What about the speed of light? We have acquired extensive knowledge that it is *the fastest anything can go*. Or should we rather say *any thing,* rather than *anything*? Is there *something*, or *some thing*, that is *not* subject to the limitations of the speed of light? We now know that the answer to this question is a resounding "yes."

In this book, I am presenting a perspective that, while the laws of physics are not violated, allows for, in fact *postulates*, that there is not only *something,* but an entire reality of *no-things* that is nonphysical.[4] To me, the significance of this *no-thing-ness* is compelling. I could no longer satisfactorily live in the confinement imposed by the limitations of the "classical" physical reality. The frame became too small.

---

[4] What I mean here is *nonphysical in the traditional sense*; see Footnote [2].

CHAPTER 1

Expansion of Energy

From Supreme-Original Energy to Consciousness

If consciousness is anywhere in the universe, it must be everywhere.
Willis Harman

Summary

==

I am presenting here (i.e., only for Chapter 1) a chapter summary to allow the reader to skip the detailed reasoning presented in the body of this chapter[5] and go directly to Chapter 2, without loss of context for the rest of the book. I must, however, admit that this summary may be somewhat difficult to read – it condenses a number of complicated concepts. It may, therefore, be more helpful for the reader who is less trained in the sciences to skip this summary and instead simply read the highlighted blips in Sections 1.1 through 1.10, before proceeding to Chapter 2.

==

The hypothesis is discussed that a counterpart to the entire physical reality exists such that this "counterpart reality" and the physical reality form a dualism. In extrapolation from the mass-wave dualism in the physical reality, the counterpart reality has an additional physical dimension, which eliminates the limitations in space and time inherent to the physical reality and provides *temporal and spatial omnipresence* in the counterpart reality over the physical reality. Arguments are presented that this additional dimension is equivalent to an understanding that all processes in the counterpart reality occur at velocities many orders of magnitude greater than the speed of light.[6] The rationalization continues with

[5] Chapter 1 is mostly an excerpt from my book "*Consciousness or Entropy?*" (Eloret Press, 1991)[18)].

[6] What this means is that, from the frame of reference of the physical reality – which is the "natural" frame of reference we humans have – processes in the

the realization that the speed of light, which is considered the *maximum* speed that can occur in the physical reality, is, in fact, the *minimum* speed occurring in the counterpart reality. In terms of velocities, the two realities are, thus, seamlessly complementary.

In analogy to the wave-particle dualism in physics, the spiritual reality (called *counterpart reality* in the first chapters of this book) is understood as the *dualism counterpart* to the entire physical reality.

Both realities have *energy* as a key element. In the physical reality, energy is an *intangible* expression with two tangible dualism aspects: mass and (electromagnetic) waves. In the counterpart reality, *energy* is just one dualism aspect of a higher form of energy. The other dualism aspect, or counterpart, of energy is *thought*, so that "energy" and "thought" in the counterpart reality are analogous to "mass" and "waves" in the physical reality. They are the major dualism elements in the respective realities.

"Energy" in the physical reality has a dualism counterpart in the spiritual reality: "Thought."

---

counterpart reality would be essentially *infinitely fast*. Also, the "physical" space taken up by the counterpart reality would be *infinitely greater* than the – already essentially infinite – space in our physical reality, making an "object" like the Earth but a small speck when seen from the vastness of the counterpart reality. This small item would be visible or perceptible from the perspective of the counterpart reality *as a whole*, in its entirety, which is not dissimilar to this book page being visible to you *as a whole*, all in one piece. Therefore, inasmuch as you and I can be "omnipresent" on a book page, the entire Earth – or even the entire physical universe, can be perceived as omnipresent from a hypothetical conscious entity in the counterpart reality.

Continuing the dualism analogy, there must be a higher-order reality that contains the two realities, physical and counterpart, as a dualism. We call this the *Supreme-Original Reality.* In as much as waves and particles are dualism aspects of *energy* in the physical reality, energy itself is only just one dualism aspect of a higher ("supreme") form of energy in the Supreme-Original Reality. We refer to it as *Supreme-Original Energy.* In the counterpart reality, "energy" and "thought" are similarly tangible as mass and waves are in the physical reality, and "Supreme-Original Energy" is similarly intangible in the counterpart reality as "energy" is intangible in the physical reality.

The physical and counterpart realities are dualism aspects of a higher-order, ultimate reality that we call *Supreme-Original Reality.*[7]

Since "energy" is an element of the counterpart reality, "entropy"[8] is also an element of that reality, and the energy-entropy process (i.e., the second law of thermodynamics) is also rooted in the counterpart reality. It is only one dualism *aspect* of a *supreme-original creative process*, the other dualism aspect being the *thought-consciousness process.* Furthermore, "disorder" and "complexity" are understood as dualism aspects of "entropy" in the physical reality, and "entropy" and "consciousness" form a dualism in the counterpart reality. Thus, "disorder" and "entropy," as well as "complexity" and "consciousness" are analogous in the physical reality and counterpart reality, respectively.

---

[7] To me, synonymous terms to "Supreme-Original Reality" are "Divine Reality," or "God."

[8] Entropy is "used" energy, energy that is not further usable for physical processes; the physical end product of physical processes.

Teilhard de Chardin's "*Thought-Consciousness Process*" is the dualism counterpart of the physical "*Energy-Entropy Process.*"[9]

From these dualisms and analogies, a dynamic energy → consciousness process is postulated. In a two-fold, unidirectional cause-and-effect relationship, an expense[10] of *Supreme-Original Energy* eventually leads to consciousness. First cause[11] is transformation of *Supreme-Original Energy* into physical energy that is usable in the physical reality. First effect is the creation[12] of conditions in the physical reality for life and ultimately physical/biological beings capable of "reflective thought" to emerge. Second cause is the thought action of man, having consciousness as potential outcome. Although first cause and second effect (i.e., consciousness, the intended outcome) are not part of the physical reality, the process depends on the physical reality for functioning.

---

[9] The energy-entropy process is a fundamental law in conventional physics (the second law of thermodynamics). It states that no energy is ever lost. All that can happen is that, as the consequence of a physical process, higher-grade energy is converted to a lower grade of energy. The lowest grade of energy is *entropy*. Once in the state of entropy, that energy has lost all its power and potential for fueling further physical processes.

[10] "expense," or use of …

[11] The "Big Bang," the beginning of our universe, was *not* first cause! First cause was what *originated, caused* that event.

[12] The word "creation" was deliberately chosen here as most appropriately fitting. Looking at the "Big Bang" with the limited frame of reference that we have in the physical reality, one might be tempted to use the word "evolution," instead. What I am really saying is that the theory of evolution, as we know it, may be entirely correct and appropriate, *but it is not the entire story*. What *seems* to us like random evolutionary occurrence was, in fact, possible only in a larger framework that not only allowed for this evolution to occur but actually thoughtfully *provided* for it to happen.

A shortcut from *Supreme-Original Energy* directly to consciousness, bypassing the role played by the physical reality, would not fit this train of thought derived from the dualism principle. Since the most noteworthy outcome of this dynamic process seems to be consciousness, the goal of the *energy → physical action → life → man → consciousness process* may well be to maximize the flow of energy toward consciousness[13].

This train of thought ultimately points to a "divinely" intended and designed process leading from the "creation" of physical energy to consciousness, man being a key factor in this process.

[13] ... and to minimize the concomitant contribution to entropy.

1.1 Dualism*[14]

===

** As hinted in the preface to the Summary of Chapter 1, Sections 1.1 through 1.10 are intended to lay a thorough foundation for the thoughts presented in this book. The reader who is less inclined to go through such detailed reasoning is encouraged to simply scan through the highlighted blips and move quickly to Chapter 2, without essential loss of context.*

===

Among the most significant scientific realizations of the 20th century ranks the discovery of the dualism principle in physics. I have described this natural law elsewhere[15] and used it as basis for the hypothesis of a *dualism counterpart reality*. In the following, I will reiterate the train of thought that had led to the hypothesis.

One might ask why we go through all the effort of discussing basic laws of physics if the objective is just to point out that dualisms exist. One might argue that dualisms can be observed everywhere. However, I make a clear distinction here between "dualisms" and "opposites." The former, synonymous with "duality," is commonly defined as a theory that considers reality to consist of two *irreducible* elements or modes.

---

[14] Sections marked with an asterisks (*) have some detailed contents that the reader who is more interested in conclusions and personal applications may wish to skip.

[15] In "Consciousness or Entropy?"18)

In the context of this book, the definition of dualism includes that the whole can be described fully with *either* of the elements or dualism aspects.

The word "opposite" is derived from the Latin word "*oppositus*" which means "diametrically positioned." Opposites are diametrically different in nature or character, they are contrary to one another, they are the other item in a matching pair. In many cases, opposites are extremes, the ends of a "sliding scale."[16] The familiar relationships "right-left," "bright-dark," "good-evil," "life-death," "masculine-feminine" are opposites.

Whereas opposites can indeed be observed all-over in nature, the occurrence of a dualism principle in the physical reality is not at all self-evident. Discovered barely a century ago and meanwhile established as key principle in physics, it is, in fact, quite surprising. A dualism characterizes one phenomenon (such as energy) as having two aspects (mass and electromagnetic waves) that seemingly have nothing in common, although the *phenomenon* (energy) is *both* mass and wave *at the same time*. Light rays are not only waves, but are particles at the same time. Electrons are waves and particles. Neutrons and protons are not only particles but also waves.

The word "dualism" or "dualistic" has a specific meaning in philosophy, which can be misleading in this context. There it refers to an understanding that life consists of two realms, physical and spiritual. The spiritual "dimension" is the place for all non-physical phenomena and beliefs. It is assumed to be out there, somewhere, separate from the physical here and now. It is often viewed as accessible only to certain, chosen people who meet spe-

---

[16] Synonyms of the word "opposite" are contrary, contradictory, antithetical, seeming irreconcilable,

cial conditions. No attempt is usually made by followers of a dualistic world view to consider the two realms as intertwined, as *two aspects of one and the same all-encompassing reality*, i.e., as what the word "dualistic" would infer to a person with a predominantly natural sciences frame of reference. In philosophy, dualism implies separateness, an either-or condition; in physics it implies two aspects, two views of *one and the same reality*; i.e., it implies a "*both and*" condition. It is important for the reader coming from the philosophical point of understanding to yield to the described new meaning of the word *dualism* as it pertains to this book.[17]

The reader coming from the philosophical point of understanding is asked to yield to the meaning of the word *dualism* as it pertains to this book: it implies *both-and*, <u>not</u> *either-or*.

One should note that some Eastern religions do, in fact, raise the three major pairs of opposites (good-evil, life-death, masculine-feminine) to the realm of dualisms. They ascribe both members of a pair of opposites to a person simultaneously. The Yin-Yang, a Chinese symbol used in Taoism, expresses that a person has the imprint, the possibility for both aspects within himself. A person, with the innate capacity of reflective thought, can consciously select one or the other of the aspects as appropriate response to any situation.

For reasons of clarity it may also be helpful to distinguish between a dualism and an "analogy," which describes a likeness, a parallelism, a similarity. When speaking of analogies, the inference is that if two or more things agree with one another in one respect, they will probably agree in other respects. For example, certain chemical elements placed in the same group of the periodic

---

[17] This is a prime example of the shortcomings of our language that no appropriate words appear to exist that would convey this difference of understanding without such lengthy clarifications.

table of elements, such as carbon and silicon, have certain physical and chemical characteristics in common. From the extensive knowledge we have about carbon and its role in organic chemistry, we can, for instance, by analogy deduce certain types of chemical bonds and molecules that could be formed with silicon rather than carbon as basis, such as "aromatic" silicon compounds or "hydrosilicons."

1.2 The Counterpart Reality*

From the law of dualisms that exists in the physical reality, which states that one entity can be described by one aspect as well as by another, I have extrapolated and formulated the following hypothesis:

> **There may exist a counterpart to the entire physical reality, such that this *counterpart reality* and the *physical reality* form a dualism.**

This hypothesis describes a new, different reality that exists simultaneously with our physical reality and is overlaid upon it, or interwoven with it. The hypothesis must be expanded with the postulation that

> **This counterpart reality and the physical reality would be the duals of an entity consisting of, and containing, both realities.**

The hypothesis, furthermore, includes that

> **The counterpart reality is *analogous* to the physical reality**

in certain respects. For example, since the physical reality contains dualisms, the counterpart reality would also be expected to contain certain dualisms.

The method of testing a hypothesis is that during the test, and only for purposes of the test, the hypothesis is assumed to be a fact. The result might then prove or disprove the hypothesis. For the present task, we work with analogies and, considering the dualism principle in the physical reality, use simple analogies and extrapolations to search for certain characteristics that the counterpart reality might have.

1.3 The Counterpart to Energy*

The dualism parent of mass and waves is *energy*. Energy is one of the most widely described and used and, at the same time, least understood concepts in physics. There is kinetic energy, potential energy, chemical energy, electrical energy, electromagnetic energy, atomic energy, nuclear energy, quantum energy, gravitational energy, heat (energy), physiological energy, free energy, and entropy ("spent" energy), to name just a few of the most commonly used types. And, of course, there is "wave energy" and "mass energy" - the all-significant dualism in our physical reality.

We know in great detail, how these energies relate to each other. We can express them in equations, we can calculate them, predict their amounts, draw conclusions from them, use them to heat our homes, propel engines, amplify sound, do calculations for us (in computers), but *we do not really understand the true nature and origin of energy* at large. Our understanding of "energy" is *phenomenological*, not *fundamental*. We excel in *describing* it, we can feel it, see it, use it, but when it comes to the understanding where it *ultimately* comes from, our concrete knowledge turns into speculation, and our confidence turns to humility.

The analogy principle, applied to the counterpart reality, would suggest that an analogue should exist in the counterpart reality to the mass-waves dualism in the physical reality. This analogue should also be a dual. In an attempt to extract further analogies from the mass-waves dualism for the characterization of that analogous dual in the counterpart reality, but without becoming too specific at this time, we note that, for our frame of reference of beings rooted in the physical reality, "mass" is the more real, the more *tangible* of these duals. We can touch (Lat. *tangere*) mass, we can form it, weigh it, eat it, look at it, whereas waves are more intangible, we usually cannot touch them (we can, at best, *perceive* them). For reasons of analogy, the counterparts to mass and waves in the counterpart reality should then also be successively more intangible, and for the frame of reference of "beings" rooted in the counterpart reality, the analogue to "mass" would be more tangible than the analogue to "waves."

Although all action within the physical reality depends on the spending of energy, energy itself is an intangible entity. Its *effect* is known and predictable; its *origin* is not. We now hypothesize:

***Energy* is an entity actually belonging to, and originating in, the counterpart reality.**

By *analogy*, "energy" would then have a dualism counterpart in that reality. We call it, for now, "Y." With this postulation of energy being an element of the counterpart reality, we have in fact established *energy as an essential link* between the counterpart and physical realities:

Energy is a key *element* in the counterpart reality and a *driving force* in the physical reality.

Following the dualism principle, "Y" is then one of two aspects of an entity that can be described by, and contains, energy but is more encompassing, just as in the physical reality *waves* are one aspect of an entity (energy itself) that can also be described by, and contains, mass[18] but is more encompassing. We call this new

---

[18] This description of energy as mass or mass as energy is, of course, the famous equation first established by Albert Einstein,[11)] **$E=mc^2$**. In this energy-mass equivalency equation, the term "c^2," which is the square of the speed of light, is a "proportionality factor," a constant, something that is in principle irrelevant to what we are discussing *here*. (However, we will see later that the speed of light, "c," is a highly relevant constant in an expanded context). The *relevant* state-

entity "X." Just as "energy" has two aspects (mass and waves) in the physical reality but itself belongs to the "next higher" reality (the counterpart reality), the entity "X" has two aspects in the counterpart reality (energy and "Y") and must itself be superior to the counterpart reality.[19] And analogous to placing the entity "energy" as an element in the next higher reality (i.e., into the counterpart reality), rather than in the physical reality, where it is a phenomenon that can be described but whose origin cannot be understood, we have to place the entity "X" in a reality ranking higher than the counterpart reality. We call this reality the "*Supreme-Original Reality.*"

***Energy* has a dualism counterpart "Y" in the counterpart reality, and *energy* and *Y* are duals of a higher-degree energy, "X," which we call "*Supreme-Original Energy,*" and which is an element of the "*Supreme-Original Reality.*"**

One might, in principle, conjecture that additional realities, each similarly more complex than the preceding one, might exist. However, there is no compelling necessity for such an extrapolation to a "*multi-layered structure of everything,*" because already the Supreme-Original Reality with its duals, the physical and counterpart realities, would be completely adequate for the conclusions I am suggesting in the following chapters.

---

ment in Einstein's equation is that mass and energy are one and the same.

[19] The analogy of "energy" and "Y" as dualism aspects in the counterpart reality will be expanded upon in subsequent subchapters.

	Physical Reality	Counterpart Reality	Comments
Dualism Aspects:	mass waves	energy "Y"	mass and energy are *tangible*, waves and "Y" are *intangible* within the respective realities
Encompassing Entity	energy	"X"	phenomena, can be described but not explained in these realities

Tab. 1.1. Main duals and encompassing entities in the two realities.

1.4 The Dimensions in the Counterpart Reality*

To learn more about the characteristics of the counterpart reality, we now examine the physical dimensions that are required to describe the aspects of mass, waves, and energy. Some major characteristics of matter (mass), such as their shape and location, are basically described with three spatial dimensions, without the need of the time dimension.[20] For example, if you take a photo of a building and another one a minute later, and give one print to one person and the other print to another viewer, both will essentially see the very same features. In this example, the passing of a short time between taking the pictures will not make a difference.

The description of a wave, however, definitely requires an additional dimension (time). Consider, for example, the wave that propagates on a perfectly calm lake after dropping a pebble in the water. The distance of the "wave front" from the location where the impact occurred clearly depends on the time elapsed.[21] If you, following our example, again take two snapshots a minute (or only a second) apart, the two viewers looking at the print they received will each see a totally different image. If you tell them, "this is how a wave looks like," and they discuss this with each other, they might get into an argument, because each will see and likely inter-

---

[20] This is, of course, only valid in *classical* physics. The fact that the relativity theory *does* introduce the time dimension for the determination of the location of a piece of matter in space has, however, no bearing on the validity of the present simplistic analogy. The specific conclusions presented here also hold in relativistic physics, as will be elaborated upon in subsequent remarks and footnotes.

[21] This example was deliberately chosen, as it describes a wave that propagates relatively slowly and is, therefore, easily detectable. The waves that are usually referred to in the mass/wave dualism propagate with the speed of light (or close to it) and cannot be "seen."

pret the "reality" of a wave differently. Clearly, time plays a major role when it comes to waves.

And, finally, the description of any physical change, which always involves expense of energy, definitely requires all four dimensions (spatial dimensions plus time).

Following the analogy principle, we can then postulate that the same behavior should be found in the counterpart reality. This would mean:

The entities/energies "Y" and "X" ***would require an additional "physical" dimension***[22]**.**

At this point, after the step has been made to introduce a new dimension for the description of the entities "Y" and "X" in the counterpart reality, it may be advisable to consider the following consequence from the relativity theory, where all four dimensions ("space-time") are used with equal emphasis for the description of physical processes. *Mass* becomes a relative quantity, relative with respect to a *viewer*. If the velocity between the mass and the viewer is high (essentially close to the speed of light), the mass, from the point of reference of the viewer, increases rapidly as that relative velocity approaches the speed of light.[23] This "relativistic correction" becomes negligible if the relative velocities are small, whereby "small" is again a relative measure.[24] For example, one

---

[22] One should note that the addition of more than one dimension, as would intuitively be attributed to a reality as supreme as the one that is often perceived as *the Mind and Origin of all that is*, is not required for working with the hypothesis of the counterpart reality and for all the conclusions drawn from this hypothesis.

[23] The "relativistic correction factor" is $(1-(v/c)^2)^{-1}$, where v is the relative velocity and c is the speed of light.

[24] Even at a relative velocity of 30km/s, which is faster than any propulsion of vehicles man has ever been able to accomplish, including space vehicles, the

can calculate that the total mass, including the relativistic correction, of an electron that has been accelerated in an electric field of approximately 510,000 volts is twice the mass of an electron that has been accelerated in a field of only 50,000 volts (or less).[25]

One might, therefore, conclude that in a "relativistic corrected" counterpart reality the entity "energy" (as dualism *counterpart* of "Y" and as dualism *aspect* of "X") also contains the new dimension that was introduced for the description of "Y" and "X." This would be in total analogy to the dimension of "time" that has its part in what we understand as "mass," even under low velocity conditions. However, the *manifestation* of this fifth dimension in "energy" would be expected to be as unrealistic, or as unnecessary, as the manifestation of the relativistic correction is (or seems to be) in the *day-to-day* low-velocity occurrences in the physical reality. But it does indicate that, if *thought* were an occurrence in the counterpart reality – which yet has to be postulated – a basis for the *physical* manifestation of "*thought power*" might exist. Since the three dimensions of space plus the dimension of time are all we have in the physical reality, the effect of the additional dimension hypothesized to exist in the counterpart reality cannot be measured or perceived "naturally," i.e., within the means of the physical reality.[26] Consequently, the aspect "Y" as well as the entity "X" that

---

relativistic correction is only 0.5%, which is for most practical purposes negligible.

[25] This particular example is relevant in electron microscopy, a technique widely used in physical, medical, and biological sciences. The higher voltage, some 500,000 volts, is close to the upper practical limit of acceleration voltage in modern instruments, while 50,000 volts is close to the practical lower limit. The change in electron mass causes significant changes in the contrast formation and radiation damage mechanism for the two types of microscopes.

[26] Compare, however, the argument introduced by Tiller[39)] and discussed later in this chapter, by which some measurements of this nature may in fact be possible. However, a necessary characteristic of such experiments has always been the direct involvement of a human observer-participant, a circumstance that impedes the reproducibility of the results.

contains both "energy" and "Y" as dualisms, cannot be expected to be measurable or *unequivocally* perceivable within the realm of the physical reality.[27]

The additional dimension of the counterpart reality is already present in the physical reality – but usually not unequivocally measurable there.

Following the train of thought developed earlier, which ascribes an ever-so-minute effect of the new dimension also to "regular" energy (in the framework of an illusive "relativity theory" applicable to the counterpart reality) one might conclude that a finite amount of mystery will remain in the explanation of obviously physical phenomena.[28] And, conversely, "unnatural" occurrences in the purely physical world, as are often anecdotally reported, can likely be "explained" on these grounds.

Furthermore, the relationship that would relate "energy" and "Y" to "X" in the counterpart reality, a relationship that would be analogous to the Einstein and Planck[29] formulas in the physical reality, contains this additional, yet unknown dimension.

A hypothesis of a new dimension in the description of an expanded reality is neither revolutionary nor new. It is, for exam-

---

[27] It is interesting to note that Carlton and Tiller,[6)] in their epochal *Subtle Energy Detector* experiments, used children as sensors, because they were less prone to mentally influence their perception than adults. (See also Tiller in ref. [39)]).

[28] Such a conclusion would be hard to refute even for the hard-nosed, yet open-minded existentialist.

[29] While Einstein's formula (see Footnote 18) establishes *mass*-energy equivalency, the German physicist Max Planck developed a *wave*-energy equivalency with his formula **E=hυ**, wherein "υ" is the frequency of the wave and "h" is known as the *Planck energy quantum*, which is again a natural constant that is no more and no less significant for the subject discussed here than the speed of light is in Einstein's equation.

ple, commonplace in mathematics to calculate with many dimensions ("n-dimensional space"), and philosophers have routinely operated in terms of other dimensions. However, the step of hypothesizing *two interwoven* realities, one of which has *a predominant additional dimension that is already dormant in the other via a common and connecting element in both realities*, is not at all self-evident.

The step of hypothesizing *two interwoven* realities, one of which has *a predominant additional dimension that is already dormant in the other* is not at all self-evident.

1.5 Example to Illustrate the Effect of an Additional Dimension

The following simple example might describe how powerful the addition of just one more dimension can be. The addition of merely one dimension brings the spectrum of possibilities that are conceivable to exist in the counterpart reality beyond any imagination.

Imagine a large, flat surface containing colored designs. An ant is crawling on that surface, trying to get to a certain point within the design. Since it can only move and see and exist in this two-dimensional space, it has no means to *rationally* get to the point of interest. It cannot *see* it. Whether it reaches its goal remains pure chance. Even if the ant would have an intellectual capacity, it would not be able to get a clue about the location of, or direction toward, the goal, because it has no capacity to "see beyond," i.e., see from above and thus orient itself with respect to the goal.

Now imagine that you are watching what is going on. You can see *at any time* where the ant is with respect to its goal, and where it is going. You can even make a reasonable judgment as to whether the ant is likely to succeed or, let's say, to fail by getting stuck in a drop of glue that it is unknowingly heading toward. Imagine further that the ant has some kind of ink on its feet that makes footprints wherever it goes. You will then be able to see and know *instantaneously* where it has been at earlier times. You will see the entire history of the ant at one glance. From the ant's point of view, you would not be subject to the dimension we call time. Your vantage point, which is characterized by *just one more* dimension[30] over what the ant has, would be *infinitely* superior.

---

[30] In this example, the additional dimension is, of course, the third spatial dimension.

It is not difficult to extrapolate such an example to the immensely superior vantage point of the counterpart reality. We think we know where we are, but in fact we may not at all be where we think we are and where we want to be. Only from a superior vantage point, such as from the vantage point of the counterpart reality, would we really be in the position to judge and see where we are. From there, a view of ourselves in perspective with others and with time would be possible, and one would see where we are heading. Imagine now that a mirror would be mounted in the space above the ant. The ant would now, by "communication" with the additional spatial dimension, be able to see where it is with respect to fellow ants and its goal. Similarly, our own best – if not the only – chance to see where we are with respect to where we want to be would be through some sort of communication with the counterpart reality. Clearly, the chances of finding the right direction would be greatly enhanced by such communication.

1.6 The Velocities in the Counterpart Reality*

The example of the ant in a 2-dimensional plane observed from a three-dimensional vantage point in the previous chapter indicates that the addition of just one additional dimension in the counterpart reality essentially inactivates the impact of time, as we perceive it with the constraints of the physical reality. An "observer" from the counterpart reality would be "omnipresent" and able to perceive occurrences in the physical reality *simultaneously*,[31] whereas we see such occurrences sequentially, because we are subject to the space and time limitations of the physical reality.

We will now discuss another perspective that also leads to the inactivation of time, as we know it, and to omnipresence in the counterpart reality. We will come to the conclusion that omnipresence and absence of the time limitations can also be explained with the following concept:

> **All processes within the counterpart reality occur at speeds far greater than the speed of light, the velocity of light being the *minimum* speed occurring in that reality. This would mean that the *real* limitation of the physical reality is the speed of light "barrier."**

The concept of a second reality that is based on the principle that all velocities in it are much faster than the speed of light

[31] It is actually more accurate to qualify this statement by saying "*essentially simultaneously*." Even in the super-high velocity environment of the counterpart reality, events *technically* do occur sequentially. However, for our point of reference, the sequencing is so rapid that they *appear* simultaneous.

has been proposed by Tiller.[32] He has actually developed a highly interesting and effective model of what I have called the counterpart reality. He introduced the concept of "negative mass" and of super-high speed magneto-electric waves that are, at speeds varying between **c** and as much as $\mathbf{c}^{30}$, the energy carriers in that reality. With his model, Tiller attempted to explain many phenomena that are commonly considered supernatural.

As I will describe below, it can further be argued that the concept of velocities far greater than **c** in the counterpart reality actually *replaces* the necessity of defining an *entirely new* dimension for the counterpart reality. Instead, the speed of light itself can be conceived of being the additional dimension we have been talking about.

The "new" dimension introduced to, and prevailing in, the counterpart reality can be conceived of being the speed of light itself.

The reasoning that leads to this hypothesis includes conclusions from the relativity theory. As was pointed out earlier, the addition of a new dimension in the counterpart reality results from considerations in the classical view of physics. In a relativistic view, the new dimension should even be considered part of the *physical* reality. However, there is very little *practical manifestation* of that fifth dimension (the speed of light) in our day-to-day lives. Except for very few physical processes, such as the propagation of visible light itself, life on the physical level takes place at very low velocities.[33] When compared to the speed of light, these

[32] William A. Tiller is Professor Emeritus from the Materials Science Department of Stanford University and has been active for three decades in this exciting research. The referenced paper[37)] is from his earlier work on this subject.

[33] Jet planes fly at about one millionth of the speed of light (c/1,000,000); satellites circle around the earth at one-thirty thousandth (c/30,000) of the speed of light.

velocities are extremely low, and the physical world can be considered *at rest* in that sense. This is the *classical* view (the view of Newton's concept of mechanics). The relativity theory becomes really significant only when the velocities approach the speed of light.

It is not directly obvious that the speed of light could be a physical "dimension." It is really an *extremely high rate of change of location*, so high that it was considered beyond reach or imagination in classical physics. In the counterpart reality, however, where everything is built on motion, **c** is a small quantity and can well be conceived of as a basic unit or dimension.

1.7 How *Fast* are Velocities *Far Greater than the Speed of Light*?*

The following two examples illustrate how fast the speed of light really is and indicate the consequences relevant to the concept of omnipresence. We perceive an image on a conventional television screen[34] as stationary, even though in reality an electron beam is scanning the picture tube at very high rates (it moves from left to right 525 times per image frame and performs 30 frames per second). For an average-size TV screen, this computes to a velocity of only one fifty-thousandth of the speed of light with which the electron beam scans the picture tube.

Sequential events that change at a rate faster than approximately 1/20 sec. cannot be perceived by the human eye as separate, individual events. We therefore perceive this TV scanning process as a steady image.[35]

Let us now assume that we are sitting in a super-supersonic transport ("SSST") vehicle, some 50,000 feet above sea level. From this elevation we can easily view an area ten miles across. Let us now assume that we take off in Vancouver, Canada, and want to see the entire United States. To do this, we would "scan" the country in coast-to-coast flights, back and forth, starting at the Canadian border, each time flying ten miles further south than before, thus working our way south until we are over Miami, Florida. It would take some 250 coast to coast flights to complete this endeavor, and the total distance that we would have covered between our takeoff in Vancouver and our landing in Miami would roughly

---

[34] This example pertains to conventional American TV sets, not to *high definition* (HD) systems that use more lines per image scan.

[35] ... or the disjointed placement of certain "moving" objects recorded in 1/20 second intervals as perfectly smooth, continuous motion.

be 800,000 million miles, and it would take us some 1500 hours to complete that journey. Let us now suppose that, rather than landing in Miami, we would go straight back north to Vancouver and start the scanning process all over again.

If we imagine ourselves flying at the speed of light, rather than in our "slow" jet plane that averages about 500 mph and takes 1500 hours for one "scan," we would be back over the same location in about five seconds. We could, for example, direct our attention to a particular car cruising along on Interstate 80 near Salt Lake City; we could also observe a pedestrian walking on a street in Buffalo, N.Y., when flying over that city, and a few moments later we could see a small airplane take off from an airport near Carrolton, Georgia. From one scan to the next, only five seconds would have gone by between sightings of our three respective objects. The objects would have moved only relatively little during that time. We could *almost* say that we are *continuously* observing these three objects, and we could almost say that we observe them *simultaneously.*

If we now increase our speed by just two more orders of magnitude,[36] i.e., if we increase our speed one hundred-fold, we would be over each object 20 times per second. We should now see them move *simultaneously* and *continuously*, just as we perceive a television image. From our human vantage point, an occurrence that is really sequential and happens at vastly different locations has become continuous and simultaneous.

We hypothesized that the new dimension is, in a classical mechanistic view of the counterpart reality, the velocity of light itself. In accordance with the analogy principle, we extrapolate from $\mathbf{v} \ll \mathbf{c}$ (i.e., velocities v are very small when compared to the speed of light), which applies to the physical reality, that one

---

[36] Compare this to the earlier mentioned postulation by Tiller that velocities greater by as much as 30 orders of magnitude may characterize events in the counterpart reality.[37)] Our example is, therefore, still extremely conservative.

should find **v** >> **c** in the classical counterpart reality. This would then explain that the counterpart reality is indeed characterized by *complete spatial omnipresence* with respect to the human capacity of perception. Since **v<<c** in the physical reality, this explains why the new dimension, **c**, really has little relevance for normal day-to-day occurrences in the physical reality, whereas it has high significance in the counterpart reality; and it explains why "**c**" is in principle also contained in the physical reality as a dimension.

1.8 The Nature of the Energy-Encompassing Entity*

It is interesting to speculate what the entity "X"[37] (that has *energy* as one of its dualism aspects) may be like. Since "X" contains (in the classical view) at least one more dimension than is needed to express physical energy, it may be, or evolve into being, vastly superior and more effective than any kind of energy that we can describe in conventional physics. We recall that in the dualism of matter and waves, the quantity *energy*, of which they are both dualism aspects, is much more dynamic and contains much more potential, power, possibility to affect change than the quantity *matter* alone or the quantity *wave* alone. In direct analogy, the quantity "X" must be understood as vastly more dynamic, fast, intricate and effective than any known form of "energy." In fact, "X" may be conceived of as so superior to physical energy that it is even superior to the limitations that might exist in the counterpart reality, where it might be understood as a phenomenon only, i.e., not really explainable. As discussed earlier, it would take a step into a second level of higher reality, which I have called the *Supreme Original Reality*, to fully explain the entity "X." And it would therefore not be unreasonable to call "X" *Supreme-Original Energy*.

Table 1.2 summarizes the main characteristics of the two dualism realities.

---

[37] See Chapter 1.3.

	Physical Reality	Counterpart Reality	Comments
Dualism Aspects:	mass waves	energy "Y"	mass and energy are tangible, waves and "Y" are intangible within the respective realities
Encompassing (i.e., higher) Entity:	energy	"X" (Supreme Original Energy)	phenomena, can be *described* but *not explained* in these realities
Dimensions and Velocities:	4-dim. $v \ll c$	5-dim. $v \gg c$	classical view relativistic view
Limitations:	space-time	temporal and spatial omnipresence	can be concluded from either the classical or the relativistic view

Tab. 1.2. Characteristics of the dualism realities.

1.9 The Law and Process of Increasing Consciousness*

We have so far said that the entity "Y," i.e., the dualism counterpart to *energy* in the counterpart reality, is (in the classical view) characterized by an additional dimension.[38] In an attempt to further describe the dualism aspect "Y," we now examine the energy → entropy process (the "entropy law"), i.e., the *Second Law of Thermodynamics*. In particular, we will entertain the idea of a dualism counterpart to this process and examine what its characteristics might be. If our hypothesis of the existence of a counterpart reality holds, and if we apply the analogy principle, such a dualism counterpart to the entropy law should exist.

The Jesuit theologian, scientist, and philosopher Pierre Teilhard de Chardin[36)] investigated this question with great depth, albeit not in the dualism context I am proposing. He reported his findings in a series of papers, the most important of which was a book entitled *The Phenomenon of Man*. His search was for a general *analogy* to the energy/entropy process, which is a task that is certainly very similar to the search for a *dualism partner* for that process.

Teilhard describes processes of *ever-increasing complexity* and *ever-increasing consciousness* that are concomitant with the

---

[38] It must be emphasized that, as was pointed out earlier and will be reiterated with varying emphasis later in this book, this additional dimension is also contained in the physical reality but has no functional relevance there. All velocities in the physical reality are limited to the speed of light, which acts quite pragmatically as an activity/influence barrier in the physical reality, whereby all velocities in the counterpart reality are *greater* than the speed of light, which is the *lower* velocity barrier in that reality. This defines the speed of light as the conceptual "boundary" between the spheres of influence/activity of the two realities.

process of ever-increasing entropy. According to this principle, every physical process, including every life (biological) process, converts energy to entropy and has as a by-product some finite increase of consciousness. The law does, however, not indicate *how much* increase of consciousness can be achieved per unit energy that ends up being converted to entropy.

Teilhard bases the development of consciousness on *reflective thought*. The emergence of life and man, having the "capacity of reflective thought," are therefore prerequisites for *the thought → consciousness process*. According to this theorem, the beginning of time, i.e., the beginning of the physical reality, was marked with a maximum of available useful energy and a minimum of consciousness and complexity; and at the end of time, all usable energy will be depleted, and a maximum of complexity and consciousness will have been established.[39]

It seems reasonable, in the framework of the discussion of the two realities, to hypothesize that Teilhard's law of increasing consciousness is the dualism counterpart (in the counterpart reality) to the law of increasing entropy.

The *energy → entropy process* and the *thought → consciousness process* form a dualism in the counterpart reality, building on, and corresponding to, the *energy -"Y" dualism*.

---

[39] Teilhard essentially refrained from considering anything other than the Earth as largest relevant system. This view is far too confining for our current understanding. Considering the vastness of the universe, and the minuteness of our planet Earth within it, the "minimum of consciousness" Teilhard talked about clearly should not imply that there was no, or only little, consciousness in the Universe at the time of its planetary emergence (i.e., some 9 billion years after the "Big Bang"). Considering the truly "universal" frame, *any* amount of consciousness man may be contributing will certainly be but a minute addition to the vastness of consciousness already present in the Universe! I am expanding on this important topic at various occasions in the following chapters, in particular Chapter 3.

As a first step in the examination of this hypothesis, we further hypothesize that:

> **"*Thought*" is what we defined as aspect "*Y*," i.e., the counterpart to *energy* in the counterpart reality. *Energy* and *thought* are a dualism.**

A preliminary check of this hypothesis with what we earlier discussed about the dualism aspect "Y" and the encompassing entity "X" (later defined as *Supreme-Original Energy*) indicates logical agreement: thought appears to clearly be beyond the physical dimensions, as it should be in order to be eligible as a counterpart to energy. Yet it *is* some sort of energy, a conclusion that is both startling and plausible. Furthermore, we can project thought *instantaneously* anywhere within the universe and into any time. This is certainly in agreement with the independence of time, which we earlier postulated as being characteristic of the counterpart reality. The speed with which a thought projection occurs seems unlimited, and place and/or time do not seem to "matter" at all. We cannot physically follow a thought projection, because we are physically subject to our space-time dimensions. But the human thought capacity does indicate that man can operate with the additional dimension that we postulated for the counterpart reality.

Regarding *Supreme-Original Energy*, it is, of course, quite reasonable to conjecture that it does contain the entity *energy*, and it is reasonable (and "thought" provoking) that conscious thought is part of the *Supreme-Original Energy*. Following the analogy principle, we can then assume that *energy* in the counterpart reality corresponds to *mass* in the physical reality, and that likewise *thought* and *waves* correspond to each other in the two realities[40] as

[40] The concept of "thought *waves*" is frequently being mentioned.

do, of course, *Supreme-Original Energy* and *energy* in the next higher levels of realities.[41]

Just as in the physical reality *waves* are the common means of transmitting physical energy,[42] it can be assumed that, in accordance with the dualism principle, *thought* is the common means of transmitting Supreme-Original Energy. The dualism analogy for the complicated physical energy transmission process, which has increased complexity of physical subject matter as outcome, would then be the transmission of Supreme-Original Energy with thought waves, whereby the "complexity of matters" in the counterpart reality would increase by a finite amount. The latter effect might then appropriately be called *consciousness*.

Just as in the physical reality *waves* are the common means of transmitting physical energy, *"Thought"* is the common means of transmitting Supreme-Original Energy.

In order to perform action or work of any kind in the physical reality, energy is required. This process is manifest in the physical reality in that energy is *irreversibly* consumed (in the classical sense) and stockpiled as entropy.[43] Since *energy* is an element in the counterpart reality,[44] it is logical to attribute also

---

[41] Furthermore, in analogy to the intangibility of *energy* in the physical reality, *Supreme-Original Energy* is also intangible in the counterpart reality, whereas mass and energy are the "tangible" aspects in the two realities, respectively (see Tab. 1.2).

[42] For example: radio waves, X-rays, ocean waves, acoustic waves, heat waves, electron waves, light waves.

[43] The reader may be reminded that entropy is defined as *used energy*, i.e., it is still *energy*, but no longer useful for anything. It is wasted energy, lowest on the scale of usefulness.

[44] See Tab. 1.2.

entropy, the state of unusable energy, to the counterpart reality. The manifestation of any action or process in the physical reality is therefore concomitant with a finite increase of entropy in the counterpart reality, i.e., the process of "action" or "work" in the physical reality is paralleled with an *"energy → entropy"* process manifest in the counterpart reality.

This would suggest that one might also postulate that dualisms also exist in terms of *processes.* Analogous to "energy" and "thought" being dualism aspects in the counterpart reality, and the "energy →entropy" process being related to "energy" as dualism aspect, there should then also be a *process* aspect corresponding to the "thought" dualism aspect. It seems thus reasonable to conjecture that

The *thought* → *consciousness* process is the dualism partner to the *energy* → *entropy* process.

The word "consciousness" (Lat. *con-scire*, to *know together*, i.e., in context) suggests that it is an outcome, a result of a process involving the mind, which works in the form of generation, projection, and analysis of "thought." The *thought → consciousness* process thus addresses the creative, intelligent aspect of Supreme-Original Energy.[45]

This is in striking analogy to the "*energy → entropy*" process that deals with the *functional* aspect of Supreme-Original Energy, and in which process entropy is an outcome (a result of expending energy). These two processes, the *energy → entropy process* and the *thought → consciousness process*, can, therefore, be understood as a dualism of what we can call the "*Supreme-Original Creative Process*" in the *Supreme-Original Reality.*

---

[45] For more on consciousness, and its distinction from knowledge, see Chapter 4.

Using the associative principle[46] we can, finally, also consider "entropy" and "consciousness" to be a dualism in the counterpart reality. It would be subordinated to what we might call "*cosmic entropy*"[47] in the Supreme-Original Reality.

For reasons of completeness, one might consider that the dualism processes *energy* → *entropy* and *thought* → *consciousness* in the counterpart reality should have analogous dualism processes in the physical reality. This might be what we commonly understand as the *process of work* and the process of *physical communication*, respectively. They are dualism aspects of "physical" action, under which one would include all "basic" physical and chemical processes (involving matter and the change thereof), while "physical communication" would be those processes that are primarily designed to *transmit* (such as transmission of data). A common end or by-product of such a process of work, unless some sort of communication is involved, is disorder. Communication is the basis for order and increased complexity in the physical reality.

This then confirms Teilhard's statement that a physical action in the physical reality has as a by-product the potential of an increase in complexity. However, whether this law extends to *any* action performed in the physical reality, or only to those involving living organisms or even humans, is subject to pointless argumentation, and I refrain here from submitting an opinion.

Complexity and consciousness would thus be *analogous* terms in the physical reality and counterpart reality, respectively. They would be the outcome of an expense of energy in the respective realities. Therefore, *an increase of consciousness may well be*

---

[46] Analogy by association; if two processes are analogous, then also key associated *elements* of these processes must be analogous.

[47] See also: "Dark Matter" (Chapter 3).

considered the by-product of the human reflective thought process, which is what we set out to examine.[48]

The sets of duality aspects in the physical and counterpart realities discussed above are summarized in Tab. 1.3.

Physical Reality	Counterpart Reality
Mass and ***Waves*** (understood as dualism of the phenomenon "***Energy***")	***Energy*** and ***Thought*** (understood as dualism of the phenomenon "***Supreme-Original Energy***")
Work Process and ***Communication Process*** (understood as dualism of "***physical action***," i.e., the phenomenological "***Energy → Entropy Process***")	***Energy → Entropy Process*** and ***Thought → Consciousness Process*** (understood as dualism of the phenomenological "***Supreme-Original Creative Process***")
Disorder and ***Complexity*** (understood as dualism of the phenomenon "***Entropy***")	***Entropy*** and ***Consciousness*** (understood as dualism of the phenomenon "***Cosmic Entropy***")

Tab. 1.3. Major dualisms in the two realities.

[48] A distinction must be made between *consciousness* and *knowledge*. For more details see Chapters 4 and 5.

1.10 The Energy → Consciousness Process*

> ***Everything science has taught me – and continues to teach me – strengthens my belief in the continuity of spiritual existence after death. Nothing disappears without a trace.***
>
> Wernher von Braun

It becomes apparent from the previous subchapters that an intricate relationship exists between the physical reality and the counterpart reality, as well as the encompassing *Supreme-Original Reality*. This relationship will now be further examined. We will in particular examine the flow of energies and see, if there is an overall *direction* that might lead to what could be construed as the purpose for the existence of the two realities.

Basis and origin is an action in the Supreme-Original Reality, where Supreme-Original Energy ("cosmic thought") is transformed into (physical) matter that is usable in the physical reality.

Actually, following the logic developed earlier in this book, the process would have been in two successive, analogous steps, one occurring in the counterpart reality and the other in the physical reality. The first step was the conversion or condensation of *Supreme-Original Energy* into physical energy. The second step was the conversion or condensation of physical energy into matter or mass.[49] The *matter* we are talking about here is the "stuff of the universe," i.e., the galaxies, stars, sun, planets, earth and moon.

---

[49] Energy and mass are equivalent; the factor c^2 in Einstein's equation $E=mc^2$ is just a proportionality constrant,[11)] see Chapter 1.3.

The *Great Cosmic/Divine Plan* appears to be a two-fold, unidirectional cause-and-effect process leading from cosmic energy to consciousness. The primary cause (*Supreme-Original Thought and Consciousness*) originates in the *Supreme-Original Reality*, leading to the creation of physical energy as prerequisite for fulfillment of the conditions for thought- and eventually consciousness producing actions or processes to take place in the physical reality.

Availability and use of *matter* in the physical reality has then led to the conditions for life and – on Earth – ultimately mankind to emerge. At the beginning point of the evolution of the universe, "thought" (as dualism counterpart to "energy") was probably of secondary significance and (perhaps) of little avail. Its *development* over *chronos*[50] time, however, may well have been the substantive purpose of the entire evolution of the universe. Finally, the *thought* action (the culmination of life in the physical reality) can have *consciousness* as outcome, manifest once again in the counterpart reality.[51]

---

[50] I am borrowing this term from Ron Roth,[29)] who fittingly makes a distinction between *Chronos* time, which is the time we are subject to in the physical reality, and *Chiros* time, which is the many orders of magnitude faster time in the spiritual realm (which I equate with what I have, up to this point in the book, called *counterpart reality*).

[51] In departure from the usual style I have adopted for this book, I have intentionally presented this difficult concept with different wordings in the text and the two highlighted blips.

The secondary cause is then an action or a process in the physical reality caused by the evolutionary outcome of the primary cause, such as by cognitive beings.

This indicates a two-fold, unidirectional cause-and-effect relationship leading from cosmic energy to consciousness. The primary cause originates in the Supreme-Original Reality. The primary effect is the fulfillment of the conditions for thought-producing actions or processes to take place in the physical reality. The entire physical evolution is part of the *effect* of this primary cause. It started with the "Big Bang," which is believed to have marked the beginning of the universe some 13 billion years ago and – with regard to our planet[52] – went through the marvelous stages of the evolution of life, from the first primitive microbes to the emergence of mankind. The secondary cause is then an action or a process in the physical reality, originated by the end-product of this evolutionary development, i.e., man. The effect of this secondary cause can be an increase in complexity or, involving thought, increased consciousness.

[52] It must be reiterated that nothing said in this book should be construed as a view that mankind might be the only species in the universe that has the capacity of reflective thought. However, it is at times more *practical* to talk about mankind and its uniqueness as if that were the case. There might well be, or have been, other intelligent life forms elsewhere in the universe. What is being expressed here pertaining to "man" is not meant to exclude those other intelligent life forms and the place they may have in the overall scheme of the evolution and purpose of the universe. The word *man*, as it is used here, should instead be understood as representative of all life forms that have the capacity of reflective thought. It has no exclusive (and, of course, no sexist) connotation.

The effect of this secondary cause can be an increase in complexity and/or, involving thought, increased consciousness.

This description indicates an intricate relationship between the *effect* in the first cycle and the *cause* in the second cycle, i.e., man. The *action* on the physical level has self-fulfilling character, in that it leads to increased complexity and, consequently, improved conditions for eventual thought "output," leading ultimately to more and more *effective* "production" of consciousness. The *physical action* bar in the graphic representation (Fig. 1.1) must, therefore, be understood as a process moving *in emphasis* from left to right, with some finite contribution to *entropy*, *complexity*, and *thought* in principle stemming from each physical action. During the course of evolution on planet Earth, the emphasis has shifted from entirely physical actions, involving inanimate matter only, to more and more emphasis on life and finally man, who became the dominant product of the entire global evolutionary process. The more to the right in the graph an action occurs, the higher is the potential to end up contributing to consciousness, rather than merely to entropy.

Primary cause (*cosmic thought*) and secondary effect (*increased consciousness*) are both manifest outside the physical reality. However, no direct line, no shortcut exists from *supreme-original energy* to *consciousness*. Instead, the "detour" through the physical reality is required.

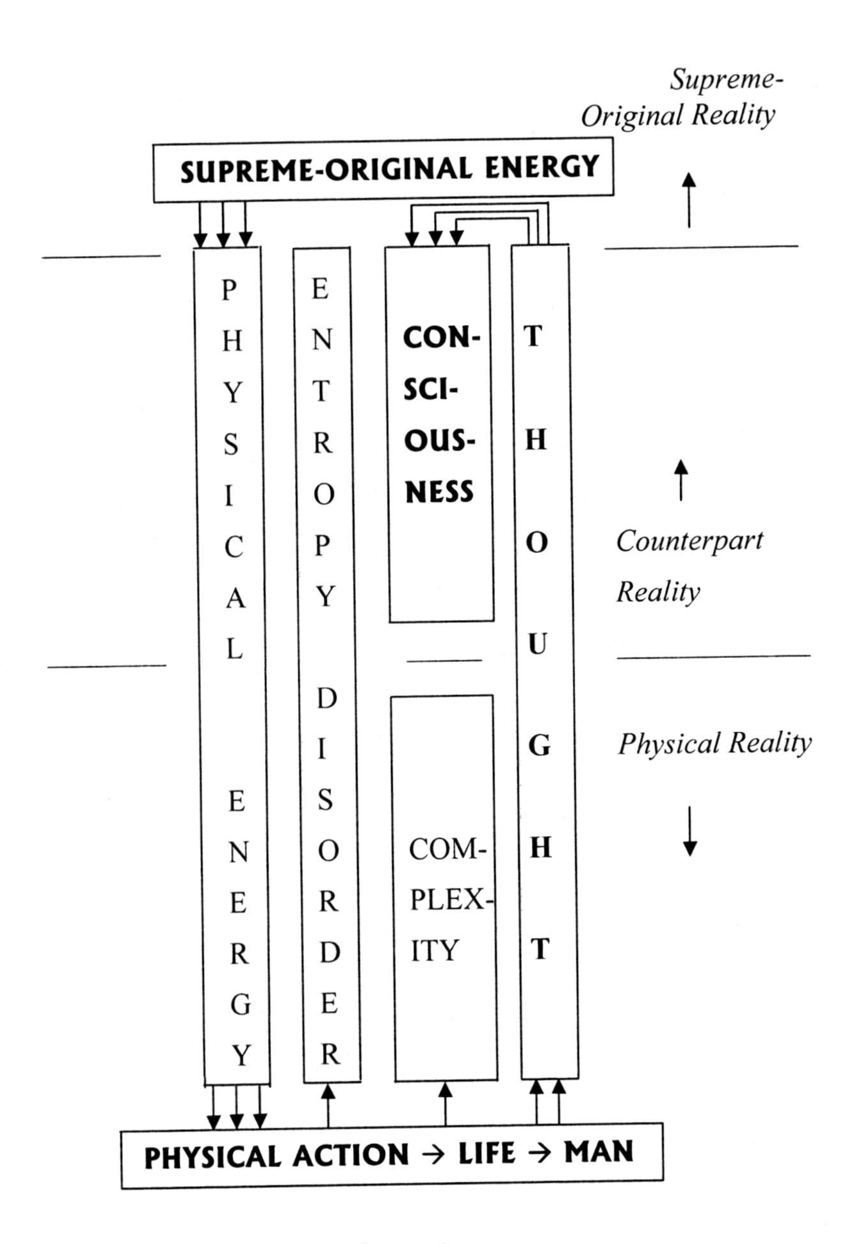

Fig.1.1. The energy → consciousness process

Primary cause (Supreme-Original Energy, which one could also call "cosmic thought") and secondary effect (which is consciousness, which is what I believe is the *intended outcome*) both take place outside the physical reality. However, no direct line exists to shortcut from *supreme-original energy* to consciousness. Instead, the "detour" through the physical reality is necessary.

The diagram of this twofold process also underscores the earlier mentioned dualisms and analogies. In addition to the dualism *physical energy* and *thought*, which are the major dualism *elements* in the counterpart reality, the dualism pairs *entropy-consciousness* and *disorder-complexity* become apparent. They are all forms of spent energy. Entropy is the form of physical energy that can no longer be used. It is the *negative* end product, the final waste.

The dualism counterpart to disorder is complexity. It represents the highest value that can be ascribed to the *physical* reality. It has evolution furthering value, in that it improves the conditions for further actions and processes in the evolutionary development of life. Complexity is, therefore, the positive aspect in the physical reality of spent energy.[53]

The equivalent *positive* end product of energy in the *counterpart* reality is consciousness. While "entropy" describes the ultimate waste, uselessness, disorder, triviality, its dualism counterpart "consciousness" describes the ultimate benefit, usefulness, order, meaning, purpose.

The arrows in the graph point out the directions in which the *energy → consciousness* flux proceeds. Those leading to en-

---

[53] Note that I am deliberately not ascribing any value in the "traditional" sense, such ethical values, to the physical realm. Love, compassion, altruism, honesty, modesty, respect, kindness, humility, meekness, chastity, whatever traditional value we might list, are non-physical values and fall in the overall classification of consciousness, rather than complexity.

tropy are based on the second law of thermodynamics. The arrows in the lower right hand side are analogous those that in the upper left and thus, in analogy, add understanding to that mysterious process of creation of physical energy at the beginning of time[54] and space. Just as *thought* is the result of an action of an aware person, the original creation of physical energy must have been the result of an action of an immensely more aware entity, the product of a Supreme Will. The arrows from the lower horizontal bar to *complexity* and from *thought* to *consciousness* are in agreement with Teilhard's *law of ever-increasing complexity and consciousness.*

The circumstance that there is no arrow indicated in Fig.1.1 from *thought* to *complexity* defines what we mean to understand as "thought" in this context: it is *not* an analytical function or capability of man (such as "learning") – which most certainly would lead to what we call complexity – but predominantly an *ethical* function of man. It is rooted in a person's inner self and expresses values and the person's respect for the incredible Oneness we are all part of.

The most noteworthy outcome obviously being consciousness, the goal of the action-life-consciousness process is believed to be to *maximize the flow of energy toward consciousness.* The hypothesis of the energy → consciousness process would apply anywhere else in the universe where life forms might exist that evolved, or have the capacity to evolve in the future[55] to the degree that reflective thought might become possible. The word "man" in Fig.1.1 should then be understood, most generally, as *life form capable of reflective thought.*

---

[54] *Chiros* time

[55] Better yet: in what we, viewed from our limited frame, would call the *future.*

CHAPTER 2

Expansion of Space

People usually consider walking on water or in thin air a miracle. But I think the real miracle is not to walk either on water or in thin air, but to walk on earth. Every day we are engaged in a miracle which we don't even recognize: a blue sky, white clouds, green leaves, the black, curious eyes of a child – our own eyes. All is a miracle.

Thich Nhat Hanh

2.1 The "God Thing"

"I am going to be doing a God-thing to this poor creature," my wife whispered into my direction. Of course, I had no clue. Then she showed me a bug, quite a sizeable critter, at least half an inch in size, floating on a tiny leaflet on the Mediterranean waters in a small sheltered cove we had selected for a relaxing swim. The poor thing had somehow found its way into this distressing situation. It was actually about to sink. Apparently it *knew* it. Its feet were already under water; it had kind of hidden them under its body, as if trying to swim or preserve energy, whichever it may have been. Clearly, the handwriting was on the wall, namely that it would entirely sink and drown in just a short while. But did it? Of course not, because Gundi did her "God-thing." She picked up the leaflet with the frightened creature, all curled up in fear of its impending demise, and carefully deposited it on dry land, where it belonged.

We started to reflect on this situation. The little bug had absolutely no idea what had happened to it. If it knew anything – and let's assume it did – it realized that it was about to die and then, for some *miraculous* reason, was airlifted onto land to safety. It was saved. God had saved it from assured death by drowning! How else could it possibly have perceived what had happened? For it, Gundi was "*God*."

What does this little happening have to do with the topic of this chapter? There is indeed a serious connection. You'll find out soon. First, however, let's realize that, of course, no *real* miracle had happened at all. May be, the miracle was that Gundi had noticed the critter to begin with, but let's set that aside for now. What's then left is an entirely plausible, calculable, predictable, physical event: Gundi lifted the leave with the bug on it, turned around and, quite within her reach, placed it on safe grounds. Nothing special about it, or was there?

Indeed, from *our* human perspective, there was nothing special about it. But from the vastly inferior perspective of the little bug, it was nothing less than a complete and immensely grand miracle. Now then, let us proceed into the chapter on "Expansion of Space."

2.2 Size Relationships and Micro-Macro Analogies

Up to this point in the book, I have used the dualism principle in physics as basis for the hypothesis that the physical reality may only be one dualism aspect of an all-encompassing, higher-degree reality. And from the dualism principle we were able to derive rationale for some of the characteristics of this (essentially nonphysical) dualism *counterpart reality*. There are, however, other striking facets of the physical reality, other interesting laws and relationships that, when suitably viewed and/or extrapolated, corroborate this hypothesis. One fascinating subject of interest has been to me looking at the micro cosmos and the macro cosmos and drawing conclusions from the striking similarities we find in these so vastly different scientific disciplines that are literally at the opposite ends of the ultimate size scale.

This interest actually goes back to my formal education as experimental physicist. My Ph.D. thesis was in high-resolution electron microscopy. I was involved with calculation of the theoretical limit of a particular type of electron microscope objective lens[56] and demonstrating that resolution in practice. I used two types of microscopes, the essential elements of which I built in the laboratory of the Institute of Applied Physics in Tübingen, Germany. My thesis advisor was the director not only of this institute but also of the Institute of Astronomy. He taught fascinating lectures in both fields and never missed pointing out the similarities of the *micro* and the *macro* worlds. Even though the body of knowledge in both fields has been greatly expanded since then, the *basic* concepts of electron microscopy, down to atomic-level image resolution, and of astronomy had already been discovered, and

[56] An electrostatic "immersion" lens, used both in "transmission" microscopy to achieve exceptionally high contrast with unstained biological specimens and in "emission" microscopy for imaging surfaces of metallurgical specimens.

most of the discoveries made since then involved refinements rather than *basic* new facts.

Let us first look at the micro-cosmos. Our physical bodies consist of some hundred billion cells that each contain what can be compared to a "factory of machines and workers" that all perform under very specific instructions, building our bodies and maintaining our health. In spite of this enormous quantity of cells, each has its place. Indeed, it's a very unique place. Some are more, some are less important. If something goes wrong in the factory, other cells come to the rescue and, in most cases, solve the problem. Each cell contains encoded information about the *entire* body, how it is put together and how it has to function. This means that each of these billions of cells has an imprint of the entire physical being of the person it is part of. None alone *is* the person; none has the entire consciousness of the person, but in aggregate they make up the oneness and uniqueness of the person. The presumably low degree of consciousness contained in each individual cell contributes its part in enabling the – for the frame of reference of the cell – presumably vastly superior consciousness potential of the person of whom it is a part.

In extrapolation, the concept of a collective, higher-degree consciousness, which is enhanced as a consequence of the consciousness contributions of individual people, as I am describing in this book, appears to be consistent with this example.

Going further down into the microscopic world, we find that each cell is composed of a huge number of a great variety of molecules. Each of them has its place in the cell factory. Each *knows* what to do and what not to do as a member of its "cell society." It may not have an imprinted consciousness of the characteristics of the entire person,[57] but it does display "consciousness" of its role within its sphere of influence, i.e., the cell.[58]

Are there parallels to be drawn? Perhaps the cell can be likened to our living planet Earth, sometimes appropriately called *Gaia*. The molecules in the cells might then be compared with us humans, and the entirety of the human body, consisting of hundred billion "*Gaia*-"-like cells, might, in that analogy, represent our galaxy. And humankind at large might then be likened to the entire Universe, composed of billions of galaxies. The numbers would come out about right:[59] there are about as many molecules in the cells as there are human beings on our living planet Earth; and about as many cells in the human body as there are planets and stars in our living galaxy with its tens of billions of member stars; and there are about as many people in our human societies as there are galaxies in the universe.[60]

---

[57] This is possibly with the exception of the DNA strand itself, which is, actually, one giant molecule.

[58] In "*Living with Joy*," the author Sanaya Roman[28)] likens this cellular consciousness as each cell having "within it the hologram of the whole of you." This wording, which was channeled from her spirit master teacher *Orin*, describes the situation quite pointedly, using to the same end a different, very fitting scientific metaphor. (See Chapter 2.4 for more on holography).

[59] I am comparing here orders of magnitudes and am not making any attempt of being overly precise.

[60] We expand on this "cosmic analogy" (*molecule → cell → human body →* people on earth vs. *person → Gaia → Milky Way Galaxy → Universe*) in Chapter 3. It is estimated that the Universe is comprised of several billion galaxies. Therefore, the potential and collective consciousness of the entire living Universe, seen as *one* conscious entity, may have a truly incomprehensible magnitude!

Compare:
Molecules → cells → human beings → people on earth
With:
Human beings → Earth → our Galaxy → Universe

Each stage has its own level of consciousness. The leap from consciousness at the cellular level to human consciousness would be comparable with the incredible increase from human to *galactic* consciousness. And there would yet be another jump of at least that same order of magnitude from the galactic to the *universe* levels of consciousness. What an incredible extrapolation for the potential of conscious consciousness!

Within the scheme of the entire Universe, our individual human consciousness is no more than the consciousness of a cell within the scheme of a living human person.

If we take this perspective, which is quite realistic, to heart, it would teach us humility. In the galactic scheme of consciousness, we really have no more consciousness than our cells have within our body.

This may sound little. But we can also look at it the other way: in comparison with the consciousness we individual humans have attained, or have the potential to develop, the consciousness of the entire Universe is billions and billions and billions times greater! It does not matter how small or insignificant you might personally prefer to judge our individual human consciousness, any number greater than zero – and human consciousness is *by definition* greater than zero – multiplied with a huge number gives a huge result!

Or let us approach this important analogy from yet another angle: getting back to the story of the bug on the leaf in the ocean at the beginning of the chapter, we must concede that the consciousness even of the bug, with millions and millions of "conscious" cells making up its identity, is undoubtedly millions and millions times greater than the consciousness of each of its cells. Yet compared to our human consciousness, that of the bug is for all practical purposes *nil.* So, in comparison, the consciousness of our galaxy must not only be likened greater by as much as ours is greater than that of the bug, but it must, in fact, be assumed to be billions of times greater than even that! And then, multiply the same thing by another billion or so to get from the galactic consciousness to that of the entire universe!

The consciousness of the entire Universe must be assumed to be *billions and billions times greater* than the consciousness of an individual conscious human being.

So much for the notion that the human is the ultimate of all consciousness! How irresponsible, and what a sign of ignorance is it for us when we harbor such a notion! Next time when you are faced with a "miraculous" occurrence in your life, be it getting well from a dreadful disease or simply coming home from a long trip safe and sound, it may not hurt to think in terms of a miraculous interference, from a reality beyond your comprehension, that may have "caused" your getting well or guided you on your trip home. It may well be the truth.

2.3 "Empty" Space and Interaction/Information Velocities

Back to our cells. There is more analogy to draw from. Even though cells contain countless numbers of molecules, these are, in turn, built with a surprisingly limited variety of building blocks. Our cells mostly consist of hydrogen, carbon, oxygen, and nitrogen atoms. Just these four types of atoms make up the vast majority or our living organism. In classical physics, each of these atoms consists of a nucleus that contains essentially its entire mass, i.e., its entire energy,[61] and a "charged cloud" around it[62] whose main function is to keep the nuclei of other atoms at proper distance.[63] The entire variety of living organisms, with all of its potential for consciousness, is thus derived from a diversity of composition of just a few types of atoms.

What is even more surprising about this is the relative *size* of these details. Compared to what is commonly understood as the "size" of an atom (usually expressed by or measured as the distance of two atoms within a molecule),[64] the size of the *nucleus* can only be described as *miniscule.* The nucleus, comprising almost the entirety of mass and energy of the atom, is so small that

---

[61] According to the energy-mass equivalency equation $E=mc^2$.

[62] Modern physics has greatly expanded on this primitive model and knows of a variety of *leptons*, *quarks*, and *mediators* as ultimate elemental building blocks. However, for the point intended here, the old model suffices and provides more clarity.

[63] This is necessary, because gravity would otherwise attract nuclei so strongly to each other that they would inseparably collapse into each other – and there would be nothing but inert mass.

[64] The distance between atoms in molecules and crystals is of the order of 10^{-9} meters.[2)] The size of the nucleus is of the order of 10^{-15} meters, i.e., it is about a million times "smaller" in linear extension, and about a *trillion* times smaller in volume.

all nuclei *of the entire Earth combined* would fit in the space of a pinhead.[65] The size of the atom compared to the nucleus is not unlike the size of the entire solar system[66] compared to the size of the Earth. This means that the atom consists *almost entirely of "empty" space.* And somewhere in that empty space, very similar to the sun within the solar system, there is, as a highly concentrated thing, the nucleus, that contains essentially all the mass, i.e., all the physical energy (see Footnote [61]) of the atom.

The atoms that make up our bodies consist of comparatively as much empty space as the stars find themselves in the "empty" space of the Universe, i.e., we are almost entirely empty space.

Through the space between the nucleus and the surrounding area making up an atom, there is a constant energy/information exchange occurring over thus *relatively* immense distances.[67] In the *micro*-world or atoms, this exchange takes place at velocities close to the speed of light. For the small dimensions inherent to the micro-world, these velocities are "adequate" in that they provide for essentially *instantaneous* interaction occurrences.[68] How-

---

[65] This is what actually happens in stars that "burn out" and collapse. While their entire mass is then concentrated in a minute space, their gravity is essentially preserved. Hence, their gravity pull becomes so localized that everything that comes close will be pulled in and never again has a chance to escape – and a *black hole* is formed.

[66] What I am referring to, in this approximate comparison, is a sphere with a radius equaling the distance between the sun and its farthest orbiting planet.

[67] Compared to the size of the nucleus, the extension of the space around it is huge (about a million times larger than the nucleus); however, in "real" terms it is little, just about one billionth of a meter (one nanometer).

[68] It would take no more than approximately 0.00000000000000001 sec (10^{-17} sec) for information to depart from one and arrive at another nucleus if it were propagating in a straight line at the speed of light.

ever, in the *macro*-world, i.e., in the world of stars and galaxies, the speed of light quickly turns out to be inadequate. The time required for an electromagnetic wave to propagate from one planet to another within our solar system at the speed of light can, for example, be as much as several *hours*.[69] It appears now inconceivable, probably as inconceivable as occurrence of super-luminous velocities would have been for the nineteenth century engineer, that the information exchange between stars or galaxies that are many light *years* apart from each other would *not* occur at velocities that would render this exchange again essentially instantaneous. This then would *necessitate* velocities of information transfer media and provisions that are *many* orders of magnitudes faster than the speed of light.[70] This, in turn, would be entirely consistent with the hy-

---

[69] Signals sent from the Mars Pathfinder down to Earth took 20 minutes to reach us. A response, such as confirmation of a course correction of the little robot vehicle roaming around on Mars a few years ago, would take at minimum 40 minutes of nothing other than "travel" time. This rendered "remote control" of that vehicle extremely cumbersome.

[70] During recent decades, actual physics *experiments* have been performed that not only indicate, but actually *prove* that super-luminous velocities do exist. These experiments include Young's famous double-slit experiment (explained later in this chapter) and the Einstein-Podolsky-Rosen thought experiment which unequivocally proved that *instantaneous* connectedness exists between particles in different places (i.e., between "space-like separated" particles). This theorem was confirmed in several experiments (in the mid 1970ies) designed to prove Bell's theorem (1964). Gary Zukav[42)] writes in his 1979 book "The Dancing Wu Li Masters":

> "*In 1975, Jack Sarfatti, a physicist, took the additional step of postulating not only that faster-than-light connections exist between space-like separated events, but also that they can be used in a controllable way to communicate. In so doing, Sarfatti, in the most dramatic way yet, elevated the Einstein-Podolsky-Rosen effect to a first principle of quantum mechanics. ... The concept of faster-than-light communication ... is as much a radical departure from current physical theory as Einstein's special theory of relativity was for the accepted physics of 1905.*"

pothesis of a dualism-based reality system[71] comprising a counterpart to our physical reality in which *all* processes occur at velocities *far greater* than the speed of light, as discussed in Chapter 1.[72]

The conclusion that the occurrence of velocities far greater than the speed of light is no longer perceived as a contradiction to basic laws of physics but must be assumed and has, in fact, been demonstrated in experimental physics (footnote [70]) is paramount. It is the cornerstone not only for my own understanding of consciousness but also for essentially the entire disciplines of spirituality, spiritual teachings, and non-conventional (including "spiritual") healing. It is for this reason that I am emphasizing this point over and over in this book. I simply would not be able to construct credence to *any* spiritual experiences in the absence of this conclusion. With it, on the other hand, credence becomes highly persuasive. It's an entirely new ballgame. Concrete manifestation of spirituality in our lives, such as through spiritual healings, has not only become *possible* but *probable*.

The existence of physical velocities far greater than the speed of light is the cornerstone for my belief in spirituality, including spiritual healing.

---

[71] Note that, as a consequence of that hypothesis, the counterpart reality is likely characterized by all actions occurring at velocities far greater than the speed of light, which would be the *lower* limit of velocities occurring there.

[72] For the sake of giving some numerical idea of the order of magnitude of velocities we are talking about, let us assume that the information exchange from star to star that are 100,000 light years apart should be occurring about as fast as that from nucleus to nucleus in a molecule (i.e., 10^{-17} sec). This would translate to somewhere around 10^{30}c, or 1,000,000,000,000,000,000,000,000,000,000 times the speed of light.

2.4 Wave Forms and Propagation Media

In the physical reality that is familiar to us, "waves" propagate in a sinusoidal form, i.e., there are ups and downs and nodal points in discrete distances. For the fastest physical waves, i.e., those "moving" at the speed of light through "empty" space, the distance between the nodal points, i.e., the "wave length," is determined by the wave's "energy" through a simple equation[73] first developed by the German physicist Max Planck early in the 20th century. The higher the energy is, the shorter is the wavelength.

It is commonly believed that, contrary to most "tangible" wave forms, such as acoustical (sound) waves or water waves, these fast light and electromagnetic waves do not require a medium within which they propagate. For example, there is nothing known to be equivalent to water or air on/in which sound waves propagate. This may, however, only be an assumption by default, as no medium has been *conceivable* that would "fill" the vacuum in which electromagnetic waves move.[74] However, if the assumption of existence of super-high velocity "magneto-electric"[75] waves in the counterpart reality holds, these waves could very well represent that "missing" medium/continuum. If the velocities of magneto-

---

[73] $E=hc/\lambda$ wherein E is the wave's energy, h the *Planck constant*, c the speed of light, and λ the wavelength. Since h and c are constants, this equation represents inverse proportionality between *energy* and *wavelength*.

[74] There has actually been persistent speculation, spearheaded by the English physicist Thomas Young, about "ghost waves" that were "guiding" photons and electromagnetic waves through the empty space. These ghost waves were mathematical constructs that "had no physical existence," primarily because they would be faster than the speed of light, which was believed "not possible."

[75] This term was first introduced by William A. Tiller[37)]

electric waves are generally many orders of magnitude[76] greater than the speed of light, it could be expected that the nodal points are very far apart.[77] Consequently, the "empty" spaces in our physical reality could be understood as interspersed with a multi-directional multitude of magneto-electric radiation (or field) lines[78] that could be quite comparable to air acting as a continuum for the propagation of acoustic waves.

Yet even this would not be sufficient to explain one of the most puzzling phenomena of the particle/wave dualism. Particles (or waves, however one might consider them) *appear to know from – or about – each other over relatively astronomical distances.* The phenomenon has been well described in the literature, for example by Gary Zukav[42)] in *"The Dancing Wu Li Masters."*[79] The example usually used is that of wave interference behind a double slit. Two light waves emitted from the same source will "meet" at relatively huge spatial separation in a plane behind the slits, depending on whether one of the slits is open or closed. It is as if each of the waves knew about the condition of the slit (open or closed) and which way the other (relatively extremely far away) wave had selected.

Similar interference mechanisms come into play in holography. A hologram is a recording, such as on a photographic plate, of the likeness of an object in such a way that information from any one point of the object is contained *not in one point* of the image, as would be the case in a conventional photograph, but

---

[76] My working hypothesis is 10 - 30 orders of magnitude (see also Footnote [72] and Tiller[37)]). However, the exact number is irrelevant for the conclusions I am drawing, as long as it exceeds approximately 6.

[77] The "wave lengths" of magneto-electric waves would be of the order of interstellar distances, i.e., millions to billions of miles.

[78] See also the discussion of neutrinos and gravity fields in Chapters 3.7 and 3.8.

[79] Bantam Books, 1979. Reading about this effect may be fascinating for anyone who did not despise high school physics.

all over on the image recording plate.[80] It is as if the recording photons (or electrons in the case of an electron hologram) would have awareness of where *each and everyone* of the other recording photons or electrons chose to impact the recording plate! This can involve comparatively *vast* distances.[81] Such awareness hints at a remarkable degree of consciousness present in the system.

Because of logical conclusions from laws in the physical reality, consciousness is interspersed in and throughout all matter.

Again, such a degree of consciousness is *interspersed in matter*, not *because of* our hypothesis of a dualism counterpart reality, but *because of logical conclusions from relationships* ("laws") *we do know exist* in the physical reality. Consciousness does not reside only in certain organic compounds, such as our cells or brains; it instead is present, at an infinitesimal level, in every building block of living tissue and, yes, even in inert matter.

It is for such considerations that I steadfastly reject all notions that the human mind might be the epitome of consciousness

---

[80] In other words, there is no point-to-point correlation between object and image in a hologram. Conversely, the hologram is also a recording of an object in such a way that *each point* of the image contains information stemming *from the entire object*. You can, for example, break a photographic plate with a hologram into pieces and find that you can reconstruct the image of the object (i.e., "see" the object) *individually* from *each* of the broken pieces. (Admittedly, if the pieces get really small, the image will be increasingly "fuzzy," but it will still show the entire object).

[81] For example, I have worked with electron holography in which the wavelength of the electrons was of the order of 1/1000 of a nanometer, i.e., 1/1000 of inter-atomic distances, while the photographic recording plate was of the order of 0.1 meter, i.e., 1,000,000,000,000 times greater! It is as if people would be able to jump *hundreds of millions of miles* into any direction, and everybody would know *precisely* where *everybody else* would be ending up!

in the universe. And it is for such reasoning that I continue to believe that there is more to human life than birth, procreation, and death, but that there is a great scheme of things, a frame within which our galaxy, our solar system, our planet with all living things on it, and ultimately you and I have a place. Together we are participants in an incredibly marvelous process, one that is impregnated with consciousness and that has the capability, potential, and destiny to evolve and augment itself – *conscious evolution of consciousness*!

I must reject all notions that the human mind might be the epitome of consciousness in the universe ...

Nevertheless, in spite of this incomprehensible inferiority of human consciousness when compared to cosmic consciousness, as I pointed out in this chapter, I consider it wrong to conclude that the effect that even an "ordinary" individual human being can and does have on consciousness at large might be *negligible.* It is certainly small, perhaps even infinitesimal, but – even so – not negligible. As I have pointed out elsewhere in this book, the "production" of consciousness, through the expression of altruistic love, is our *individual and collective purpose as human beings and human societies.* Along with this purpose comes the realization that, as we evolve in our consciousness, so are we – ever so small and in ever so tiny steps – participating in the conscious evolution of nothing less than the mind of God. In as much as micro and macro cosmos co-exist, so does the validity of this analogy.

... yet it is also wrong to conclude that the effect an individual human being has on cosmic consciousness might be negligible.

2.5 Reviewing the Past

In Chapter 1.7 I have given an example illustrating that having domain over velocities far greater than the speed of light would provide spatial omnipresence. In the example of the *super super-sonic transport* ("SSST") vehicle, the viewer essentially experienced independence of spatial limitations as a direct consequence of the assumed super-high velocities. He was able to see vast areas in one sweep and observe countless occurrences in those areas simultaneously. We have not yet demonstrated the analogy to also viewing history.

An explanation of temporal omnipresence would require that we add another degree of freedom to our SSST: a change of vertical distance from the object. The information about the history of our test objects, such as how the little airplane that we observed taking off got fueled for its flight, has been spreading out into space at a velocity less than or equal to the speed of light. This is so, because this process is strictly limited to the laws and restrictions of the physical reality in which the *fastest* speed that can ever occur is the speed of light.

With a "vehicle" capable of flying substantially faster than the speed of light into any direction, as we had assumed our SSST to be, and with the proper information receiving and data processing equipment, we would be able to *recapture* and decode that entire information. We would therefore be able to *bridge over any period of (physical) time* and be, for a reference point in the physical reality, *independent of time.*

Viewing into the past is entirely plausible from a vantage point that includes domain over velocities far greater than the speed of light, such as from the counterpart reality.

In her earlier referenced book "Living with Joy," Sanaya Roman,[28)] a psychic who channeled a wisdom teacher whom she called *Orin*, presents a remark that became important to me in this context. Orin stated that it would be highly desirable if we would be able to see, as he can, each challenge or opportunity that presents itself in our lives in the context of our entire life's theme and purpose. Orin describes and, indeed, appears to have the omnipresence capability described above. With regard to the remark itself, I would even go one step further. As we will develop in more detail later in this book, our mind functions in the counterpart (spiritual) reality. Even we "ordinary" human beings have that very same capacity of viewing over our lives and drawing conclusions, at any time we want to and submit ourselves to the discipline to do so. This is, in fact, the core birthright we all have. It is part of the *Divine Plan* for us to have that capacity. *Acting on it* is the key! Yes, Orin, we understand what you are saying. Every experience placed before us is there for a reason. *You* can see it *instantaneously*, because your level of consciousness substantially exceeds ours. But even with our more limited consciousness, we are able to see the same. It's just a bit more difficult for us – but then, that's part of the Great Plan! When we have learned to make this connection between personal experience and purpose, we will have attained wisdom.

When we have learned to make the connection between our personal experience and our purpose, we will have attained wisdom.

2.6 Viewing into the Future

In principle, not only the past, but also the *future* would be open to us in the super-high speed scenario. If we had the proper – albeit certainly more sophisticated – information processing capabilities, we could, knowing *all pertinent* information from the past, essentially predict what will likely happen in the future.

With the super-high speed information exchange characteristics prevailing in the counterpart reality, and with proper information processing capabilities, we could, knowing all pertinent information from the past, essentially predict what will likely happen in the future.

It is not dissimilar to the observer of the ant crawling in a two-dimensional plane, with ink on its feet that marked wherever it has been, as I described in Chapter 1.5. The observer, who has domain over an additional dimension, has complete, instantaneous knowledge of the entire history of the creature. With this knowledge, he can make quite a reasonable "guess" about the ant's future. He could predict how long it will likely survive, because it hasn't found food for a while and is – unknowingly – far away from anything to eat or drink. He could forecast that it will likely get stuck at a nearby patch of glue, which would end its life prematurely, or whatever the situation might be. Similarly, an "observer" from the *counterpart reality*, who has a vantage point that is beyond our physical space-time limitations, would be able to see our entire life history, including our health situation and the milieu we find ourselves in, and could use all that information to come up with quite reasonable predictions about our future.

It would be questionable, however, to what extent also the future of events concerning humans could be predictable.

Such predictions would, however, more likely hold for predominantly physical, including evolutionary, processes[82], rather than those in which a conscious mind is involved. It would be questionable, and certainly open to intense speculation, to what extent also the future of events concerning humans could be predictable.[83]

One of the most spectacular authentic predictions of future events that I know of has been described in the book "*The Bible Code*."[8)] These predictions of political and societal events that have recently occurred are *scientifically proven* to be over 3000 years old and have stunning accuracy.[84] It is conceivable that, even dating back over such extremely long periods of time, *precise* knowledge of the entire human/world history up to that point in time 3000 years ago, combined with rigorous analytical – and certainly super-human – intelligence and foresight might lead to predictions of disasters, such as the attack on the Twin Towers, 3000 years

---

[82] I.e., all processes subject to the cause-and-effect law, which is in essence nothing but a different wording of the *First Law of Thermodynamics*, explaining that no work or energy can be spent without a proper adaptation of all other components in the system (which is, in principle, as big as the entire universe) so that the sum of all internal and external energies remains the same.

[83] One might, for example, argue that our decisions to do or not do certain actions are based on experience. If we make all decisions based upon experience or logic, many outcomes might well be predictable. However, we humans are known to, from time to time, defy all common logic....

[84] For example, the 9/11/2001 *Twin Towers* tragedy was forecast in *The Bible Code*, along with the *names* of contemporary political leaders, including Presidents Bush, Arafat, and Rabin.

down the time track. However, I do believe that *human consciousness* could have prevented the attack – *not knowledge, but consciousness*, conscious action upon a choice for the good, an action that would have been rooted entirely in free choice and would thus have been "unpredictable" even to such super-human intelligence. For the same reason, I do believe that other dire predictions of events yet to come and reportedly also embedded as ancient code in the Torah,[85] *do not necessarily have to* come to pass, regardless how precise they appear to be and how accurate those predictions were that did come to pass. But we should better understand *they will likely occur if we do not change our attitude.* The predictions are based on extrapolations of existing paradigms. Nothing less than a paradigm change with regard to how we resolve conflicts appears to be required.

It is for the same reason that the religious-philosophical concept of everything being predetermined *cannot be upheld.* In fact, predetermination would, as I see it, contradict the entire purpose of our existence. Achieving consciousness, which is what I see as our purpose, is predicated on the availability and exercise of free choice. If I have *no choice but* to love, I will, of course, end up loving, and that love is then instinctual and, hence, of much lower ultimate value than love exercised on the basis of entirely free choice.

Predetermination would, as I see it, contradict the entire purpose of our existence.

[85] One such prediction in "*The Bible Code*" is a nuclear holocaust in the year 2006.

CHAPTER 3

Expansion of Perception

3.1 Narrowing the Gap Between Science and Spirituality

When I embarked on my formal higher education in natural sciences in the early sixties, the gap between science and spirituality was quite widespread and deeply preoccupied me. I found it exceedingly difficult to uphold my inner conviction that there is a spiritual side to life, while professing to be a contemporary scholar of "exact" sciences and technological achievements. My academic studies of physics would not change that situation. I was preoccupied with *learning*, studying, and was unable and unwilling to assimilate at a deeper level what I had learned. Quantum physics, relativity theory, the dualism principle, were theoretical concepts with no relevance to the meaning of life.

I was unable to even *entertain* the "quantum *leap*" from admitting that science *allows* for a leap from existentialism to spirituality. For another 20 years I felt that, as a contemporary scientist, I could not readily "admit" to scientific colleagues that I believed in a life beyond the physical life.

I must admit, though, that part of the problem had been for me a certain degree of confusion between *spirituality* and *religion*, or religious practice. Having grown up in a fundamentalist Christian home, I identified my personal *experience* of Christian belief with spirituality. Most fundamentally, certain religious practices, such as prayer, attending church, or participating in religious ceremonies such as communion, were synonymous with spirituality. I was lost in a confusion between *form* and *content.*

That narrow perception was quickly expanded after my immigration to the United States in 1969. Numerous people became spiritual teachers on my journey through life. Most of them did not represent a formal religion or religious denomination. They were just *spiritual* people, persons who had a deep sense of what is *really* important in life and knew how to separate it from superficialities and trivialities.

Modern physics has provided the basis for essentially eliminating the gap between science and spirituality.

It is now established that modern physics has provided the basis for quite essentially *narrowing* - if not eliminating - this gap between science and spirituality. For me, it took several more years of "digesting," of *experiencing* the consequences of quantum physics, before I came full circle and was able to state that modern physics has not only narrowed the gap between science and spirituality but actually *suggests* the existence of a spiritual dimension. I now submit that it is a *logical consequence* of modern understanding of the laws and intricacies of the physical reality to expect that a spiritual reality exists and is as much a part of our very existence as the physical reality – if not even more so.[86]

[86] In today's abundant literature on spirituality, the opinion is often presented that just the opposite is true. They suggest that what we experience in our day-to-day lives is essentially only *secondary* (some even go as far as calling it *illusionary*), and the spiritual reality is the true, primary, much more powerful reality. Coming from an entirely different angle, I now tend to agree with this representation.

3.2 Spirituality and Spiritual Reality

The most common – yet likely the least conceded – reason for the rift between science and spirituality has been that either side would accuse the other of not understanding, or *not wanting to understand* their point of view. They would cite extreme sample cases and represent them as being the norm. For example, scientists would use the views and beliefs held by a small minority of religious fundamentalists[87] to justify their point that spirituality contradicts "common sense" rationale, not admitting that the views they are opposing are not at all mainstream opinions among those who believe in spirituality. And vice versa, spiritually inclined individuals frequently refer to opinions of certain atheist technocrats[88] and extrapolate them to be representative of the entire scientific community. Much learning has still to be done for both extremes to concede goodwill and sound reasoning to the other side.

A few years ago, my wife and I participated in the fabulous experience of *swimming with the dolphins* in the open coastal waters off the Big Island of Hawaii. In the framework of a 5-day "Dolphin Spirit" seminar, the experience consisted of swimming/snorkeling activities in the mornings and spiritual/meditative group/seminar experiences in the afternoons and evenings. Both were led by a dedicated and talented couple who had "taken up the call" to do this work from their teacher who had spiritual experiences in conjunction with dolphins and wrote two insightful books on the subject.[89] For a person tuned into spirituality and new-age

---

[87] For example, the virgin birth doctrine, the theorem of Christ's resurrection as a physical being, etc.

[88] Such as the view that human reasoning is the highest form of consciousness that exists and that nothing exists, anywhere, that cannot, either now or in the future, be fully explained with the laws of physics.

[89] Joan Ocean[25)], author of "Dolphin Connection" and "Dolphins into the Future."

type language, their approach would have appeared congenial. From those who are more reality-oriented, like myself, who have a somewhat harder time dealing with concepts like a "beaming a pink bubble of love," more tolerance is required. It took me a while to translate into language I could understand and consent to concepts like "asking permission of Madam Pele[90] to enter the waters of the sacred bay," "beaming love to the dolphins and asking their permission, if it is their will, to join us swimming," "moving into third-dimensional reality," or "tuning to higher-frequency perception." Vice versa, it took our seminar leaders a while to tolerate my attempts of "translating" their expressions into a for me more acceptable language. Both sides could see that, without this level of tolerance and the willingness to take in the other frame, we would have, at best, simply talked past each other and, at worst, labeled the other as off-base and discarded their respective message entirely.

What is spirituality? For the purposes of this writing, I define it as *a person's state of inner connection with the counterpart reality*, whereby the counterpart reality is for me synonymous with "spiritual reality" or "gnostic reality."[91] Spirituality is an attitude, an inclination, a pervasive internalized awareness that there is more to life than what a person can see, hear, touch, smell, sense, calculate, rationalize. Spirituality is knowing that there is a *conscious consciousness* which is superior to any human thought or consciousness capacity. Spirituality is seeing a poppy and *experiencing* – not just intellectually agreeing with – that it and I are one; it is the feeling of exuberance after giving birth to an idea that I *know* came about by inspiration; it is the sadness I experience when seeing a three-year old stricken by AIDS. Spirituality is wishing more than anything in this world that a loved one chooses the path of a meaningful life. Spirituality is being a joyous person

---

[90] A Hawaiian mythological figure/goddess.

[91] This is the term I used in "*Consciousness or Entropy*?"

living *in a state of* prayer. Spirituality is – to connect to the dolphin experience – seeing the uniqueness of these wonderful sea-mammals and accepting that they have something valuable to teach us, i.e., knowing that we humans are not the ultimate in knowledge and consciousness.

Spirituality is action upon the inner knowing that there is a *conscious consciousness* that is superior to any human thought or consciousness capacity. It is not tied to religious practices, rites, or places.

Spirituality is, for me, not tied to religious practices, rites, or places. However, I can *experience* spiritual connection more easily when I am meditating in a dignified place or situation than when engulfed in mundane business. Connection with the spiritual reality can occur through a number of mechanisms that have in common that they relate to the thought/feeling capacity of a person. For some people this may be through meditation, or through reading or listening to poetry, for others it may be through observing a masterpiece painting, listening to great music, or soaking in the sounds and fragrances of nature. It depends entirely on the personality type. For me, there are certain orchestral symphonic pieces[92] that, for whatever reason, have the power to create that connection more than anything else. Beethoven's *Eroica*,[93] certain

---

[92] See, for example, the "Mozart effect."[23)] It was recently scientifically confirmed that certain meditative music, such as the second movements of traditional classical symphonies and concertos (which are typically written as *largo*, *andante*, *andante cantabile,* following a fast first movement and followed by a *menuetto* or other type of semi-fast third movement in the case of symphonies and a *vivace* as third and final movement of solo instrument and orchestra concertos) have a marked healing effect on the human listener.

[93] Ludwig van Beethoven's 3rd Symphony, in e-flat major, composed in 1812.

Bach cantatas, or *Taizé*[94] chants instantaneously grab me this way, while poetry, for example, would tend to leave me untouched.[95] There is no good or bad "mechanism," what is needed is simply the willingness to find out what works and the persistence to subject oneself to those situations. In fact, the world would be boring if we all had the same tastes and likings. It is the incredible variety of personal expression that makes it interesting.

The realization of value of individual expression comes to me most readily when on travel. The experience of seeing people living lives so entirely different from what I am accustomed to and have come to like is to me the invaluable reward of traveling. Perhaps we should come to a point of admission that, unless we understand the value people experience when drinking the sacred healing waters in *Lourdes*, when they place small trays with real food on the altars of their Gods in *Bali*, or when they fall on their knees in *Istanbul* in unison at certain times of the day to pray to their God, we have not internalized the true meaning of spirituality.

Although spirituality comes natural to our species, we have the capacity and, in our western culture, widely the inclination to suppress it. This is what distinguishes us from other species. But this is also what makes it more rewarding, because the by-product of spirituality is *conscious consciousness*, which to attain is tied to the purpose of human existence.

I cannot over-emphasize this point that, to my distress, is often not shared by my environmentalist friends. As I have presented in detail in "*Consciousness or Entropy*?," "producing" conscious consciousness (which is *altruistic love*) is the ultimate purpose of the human being. There is a difference between the love automatically or instinctively expended in the animal world and genuine, altruistic love of a human being toward one another (or toward nature). The latter is based on a *conscious decision* that

---

[94] From the ecumenical brotherhood Taizé in central France.

[95] With notable exceptions, of course. For example, an evening with the poet David Wright is an experience I would not want to miss.

places the object of the love as high as oneself. It is, quite frequently, a decision taken at one's own perceived disadvantage. Yes, I do value the action of a *Compassionate Samaritan* higher than that of a penguin who sacrifices its life for the survival of its species. It has no *choice*; the Samaritan does.

3.3 A Large Enough Frame*[96]

There is no place to which we could flee from God which is outside God.
Paul Tillich

==

** Sections 3.3 through 3.9 deal with rather complicated subject matter and are intended to complement the foundation for the thoughts presented in this book. The reader who is less interested in this detail is encouraged to simply scan through the highlighted blips and move quickly to Chapter 3.10, without essential loss of overall context.*

==

Modern science has provided us with a fairly detailed basis for the understanding of the Universe. Most importantly in this context, the "Big Bang" theory of the origin of the universe has by now been widely accepted.[97] We know that over 99% of the mass of the Universe is "dark matter,"[98, 99] a finding that we will draw upon in a subsequent subchapter.

---

[96] Sections 3.3 – 3.6 contain valued contributions of my friend Ben Young.

[97] Doubters of the Big Bang/evolutionary theories are often entrenched in certain dogmatic religious beliefs.

[98] Dark Matter is contained in "black holes," i.e., it is physical matter that has burned itself out, lost all its useful energy and cooled down to near zero temperature, such that the electron charge density cloud around the individual atom nuclei could no longer function to overcome the gravitational pull of the nuclei, which allowed them to collapse into each other, thus eliminating essentially all the space that was formerly occupied by the atoms, while maintaining essentially the entire gravitational force of the matter (which is independent of the

We now know so much about the Universe that we have, for example, been able to detect an asteroid of 1-mile diameter, moving on a near-collision course with the Earth but *still a 30 years' journey away* from us. In fact, we know so much that we can state with *certainty*, based on scientific measurements and calculations[41)] that this asteroid will *narrowly miss and not hit* us (leaving present-day mankind markedly luckier than "dinosaurkind" some 60 million years ago when they became extinct due to an asteroid impact[100]).

Virtually our *entire* understanding of the Cosmos has been formed with a science that excludes such fields as physiology, psychology, sociology, and cosmology/ spirituality. This viewing frame is too narrow and in need of expansion. It is too confining.

When we examine what "*science*" means in this context of "scientific exploration of the Universe," we find that it virtually exclusively denotes an exploration "*by means of research in the physical/natural sciences*." Virtually our *entire* understanding of the Universe has been formed with a science that excludes such important academic disciplines as physiology, psychology, sociol-

---

electromagnetic properties of matter and solely a function of mass and distance).

[99] The number given (99%) does not consider a recently proposed rationale that attributes a significant fraction of the mass of the universe, perhaps as much as 50%, to neutrinos (see footnotes [107] and [127] on this subject).

[100] It is believed that the impact of an asteroid caused such a massive release of dust particles and smoke into the atmosphere that sun rays were blocked out and could not reach the ground, and darkness prevailed on most of our planet's land and water surfaces for as long as several years, which changed the environment so extremely that the dinosaurs – and presumably many other species – could not adapt and became extinct.

ogy, and cosmology/ spirituality, i.e., fields that are clearly recognized as having the status of "science" no less than the field of physics/natural sciences does. Our "viewing frame" has been to discover the *physical laws* of the Universe based on derivation, experimentation, and interpretation of basic laws of physics/natural sciences.[101]

This viewing frame is too narrow and in need of expansion. It is too confining. Imagine how it would be if our approach to learning about the human body (how it functions, how it can be kept healthy, etc.) had been confined to nothing but conventional physics and chemistry. We would then have little more than a fairly complete description of our body's *composition* and *physical and chemical properties*, but we would know little or nothing about procreation, the function of the DNA, diseases, psychology, feelings, behavior patterns, social patterns, and alike. Similarly, the conventional physics/natural sciences viewing frame does not allow for *interrelationships* that must undoubtedly exist also in the Universe at large, i.e., between the stars and the galaxies.

---

[101] John Gribbin's book "*Space: Our Final Frontier*"[16)] gives an authoritative overview of the theories and experiments that have led to the current understanding in astronomy and about the origin of the Universe. (See also his book "*In the Beginning*").

3.4 The Gaia Universe*

A big step in the right direction has been the Gaia-Earth concept, which was fittingly not "discovered" – which has been the "traditional" way of coming up with innovations in the physical sciences – but emerged/evolved and surfaced in the 1990ies when the time was right. It was the result of *intuitive perception*, rather than rationalization. The Gaia concept describes the Earth as *one living organism*, made up of many parts that, similar to the cells in the human body, all have distinct and discrete purpose and function within the organism (i.e., Gaia) they are part of.

The Gaia concept describes the Earth as *one living organism.* Similarly, the entire Universe is *one* incredibly large, incomprehensibly intricate *Cosmic Organism.*

If one extrapolates the Gaia concept further out and extends it to the entire universe, one fathoms *one* incredibly large, incomprehensibly intricate *Cosmic Organism.* Let us compare the galaxies (the "societies" of galaxies) with the human societies. Any one galaxy would then be comparable with one human being, and the stars and planets within any one galaxy would be like the individual cells in human beings.[102] The respective number pairs (of galaxies and people on Earth, and of stars in a galaxy and cells in our body) come out to be quite close.

If one accepts this "expanded Gaia hypothesis" that the entire Universe[103] is *One Living Organism* which is, in some sense,

---

[102] See Chapter 2.2 of this book, in which this analogy is further expanded.

[103] Capitalization denotes, here and elsewhere in this book, reverence to the

conscious, i.e., self-organizing, self-selecting, self-regulating and *knowing*, then the viewing frame for discovering the true nature of this *One Living Organism* should not be solely based on conventional physics/natural sciences but would have to include the conceptual framework of all of the other important scientific disciplines, including, physiology, psychology, sociology, and cosmology and spirituality.

If one accepts the "expanded Gaia hypothesis," the true nature of this *One Living Organism* should not be solely researched with conventional physics and natural sciences but also with all the other important scientific fields, including physiology, psychology, sociology, cosmology, and spirituality.

In fact, in as much as conventional physics and chemistry are hardly the major scientific disciplines when it comes to healing the human body or furthering the human mind, these other scientific disciplines would have to receive preponderance as viewing frame for discovering the nature and function of the Universe. Thus, it is essential to begin the process of translating and bridging between the old, limited viewing frame (based on conventional physics and natural sciences) and the required *expanded frame*.

Looking at the Universe with the expanded viewing frame is not *just a bit* different from what we have been doing, but it is *radically different*! Can we *really* understand, with all our being – and not just as a theoretical concept that leaves our feelings and intuition unaffected – that the Earth is *one living organism*? It is a topic worth thinking about! And then, assuming one can somehow grasp the magnitude of this Gaia-Earth concept, can one even *begin* to sense the implications of the Gaia-*Universe* concept? Can

---

greatness of what is addressed.

one imagine an *interactive relationship* between the Earth, the other planets, and the sun, or between our solar system and other celestial bodies? It will help if you step outside in a clear night and let yourself be absorbed by the awe of the moon, the stars, the Milky Way. Have you *ever* looked at them in awareness that there is a *relationship* between *you* and *all of* them?[104]

There is a *relationship* between yourself and the stars in the universe, just as there is a relationship between the cells in your body and yourself.

For me, probably due to my training as a physicist, the notion that there is a *relationship* between myself and the stars in the universe has been entirely uncongenial. To this day, I find myself considering the stars as – albeit fascinating – *objects*, such as a piece of stone is an object with which I would not normally *feel* a connection. The expanded viewing frame, based on *physiology, psychology, sociology, and cosmology sciences* terms, is still a mental exercise for me, far from having reached preponderance.

Personalizing and internalizing the expanded viewing frame is not an easy task. To get a grasp of how radically new and difficult this new perspective is, one might want to imagine how difficult it would be for an individual molecule in our body to understand the intended function, potential of unfolding, and wisdom and acquired consciousness of the entire body in which it exists! Even if this molecule were a stem cell DNA strand, with the codified imprint of the entire body, it would have but little conscious-

---

[104] In fact, conventional physics seems to have *taught* us that there should be no such relationship. The limitation we found in the speed of light, i.e. that it is the fastest conceivable speed for any communication, has given us the erroneous conclusion that physical bodies that are light years away from us, or that are moving away from us at, or close to, the speed of light could *not possibly* be in any way affected or influenced by humans, because no signal from us could ever reach them and vice versa.

ness, if any at all,[105] of its *context*, which is being part of a thinking, conscious human being.

Imagine how much greater the consciousness of an entire galaxy would be than that of an individual human being. And then again, compare the consciousness of the hundreds of billions of galaxies that exist in the universe to that of one single galaxy!

We can now build on the analogy discussed earlier between the galaxies and humanity. It appears to be an accepted theory among biologists that the process of *cognition* operates on all levels of life, even if no brain cells are involved. There is a certain level of cognition (knowledge, awareness, discernment) already in an individual cell. It "knows" what its place is within the body it lives in, and what its correct way of functioning is. As cells combine to higher levels of complexity, the level of cognition increases.[106] When it gets to the level of the human being, cognition has expanded to the degree that conscious consciousness has a chance to emerge. We have to assume that this expansion continues when we get to the next higher levels of complexity, i.e., the Gaia Earth and, ultimately, Gaia Universe levels. Imagine how much superior the consciousness attainable by a person is when compared to that of a single cell, or that of mankind at large when compared to that of an individual person. Now try to extrapolate this enormous difference into the next cosmic level: imagine how much greater the consciousness of an entire galaxy would be than

---

[105] As I noted in Chapter 2.2, a marvelous imprinted cognition does exist with regard to its *physical context*, i.e., to the physical body. Whether this includes awareness of the circumstance that this "context" is a thinking, consciousness-producing entity is what I am less sure about.

[106] This occurs synergistically, i.e., at rates greater than what the sum of the elements would suggest.

that of an individual human being. And then again, compare the consciousness of the hundreds of billions of galaxies that exist in the Universe to that of one single galaxy!

That's what you would expect as Gaia-Universe consciousness. And all of this consciousness was seeded at the very beginning, contained in that minute, less than single atom-size original seed that exploded into the fireball that evolved the Universe!

These thoughts are intended to expand the reader's frame of perception and to instill that there are many layers of consciousness in the universe. Our human one, albeit large compared to anything else we know (pertaining, at least, to our solar system), is but small compared with what is out there at large!

3.5 Dark Matter*

According to contemporary theories and analyses of scientific data obtained with the Hubble telescope and other cosmic probes, *dark matter* constitutes *almost* the *entirety*[107] of the physical reality.[108]

***Dark Matter* constitutes almost the *entirety* of the physical reality.**

What is dark matter? Whereas the Big Bang can be considered the *birthing process* of the physical reality, which very initially consisted of 100% "*live*" *matter*, i.e., matter that has the potential to fuel processes in the physical reality, live matter turns into dark matter once all potential to fuel physical processes has been used up.[109] Black Holes are the "repositories" of dark matter.[110]

---

[107] A figure of 99% appears to be the consensus among scientists specializing in this field of research (see, for example Gribbins, "*In the Beginning*"[16)]). However, this figure does not include recent findings that *neutrinos* may take up as much as one-half of the entire mass of the Universe. But this circumstance has no impact on the considerations presented here, as – for the sake of this particular discussion – the entirety of the neutrinos in the universe can be categorized as "dark matter," as well. One might include neutrinos in the energy category of "entropy," which exists in the counterpart reality.

[108] I.e., of what involves "matter" in the broadest sense of physical meaning of this word.

[109] This is equivalent to having reached the state of maximum entropy.

[110] Note the obvious distinction from how one has "traditionally" described black holes: the ultimate finality, the end of all ends!

Dark matter is entirely inert, it has seemingly no potential of unfolding or bringing alive anything that can be conceived of as *physically*[111] useful. It is motionless, without the dynamics of electromagnetic forces,[112] and has no effect other than the undiminished force of gravity,[113] which I address more closely in a subsequent chapter in this book.

When dark matter agglomerates by attracting more and more of its kind, its gravitational pull can become extremely strong. Eventually, *everything* that comes close gets brutally sucked into it and can never escape.[114] This even applies to light

---

[111] Note the emphasis on *physical* usefulness. As we will see, there may well be an immense usefulness in other than mere physical aspects.

[112] Electromagnetic forces are fundamental to the dynamism of the physical universe. Without them, physical "matter" literally collapses into itself, the – vast – spaces between the nuclei of atoms disappear, and what is left are essentially the preserved "masses" of the nuclei, in *extreme* close proximity and thus at essentially *infinitely high mass density.*

[113] Since the physical mass is (essentially) preserved in dark matter, the basis for gravity (which is proportional to mass), is maintained. Gravity is proportional to the masses involved and *inversely proportional* to the *square* of the distance between the gravitating masses. Since this distance has collapsed to almost zero within dark matter, gravity near a black hole becomes so immense that *nothing*, not even light, can escape its pull, and whatever gets close gets sucked into it and has no chance to ever re-emerge and/or escape. It is the "ultimate one-way street." "Black"-ness is the physical description of nothingness, hence the term "Black Hole."

[114] Some notable physicists consider this to be a violation of the first law of thermodynamics, which they interpret as having to include information. Accordingly, they postulate that there should be something "on the other side" of a black hole, something that allows a new birthing process – perhaps a new "big bang" creating yet another, altogether new physical reality. This view may be a consolation to those who might fear that black holes might also "gobble up" and forever destroy "consciousness." I find this interpretation unnecessary. Consciousness is not *the same as* information – it is more *likely the result of, the conclusion drawn from* information. Whereas information is a *physical* phenomenon, consciousness is not. Information can be reduced to electromagnetic waves that are subject to the mass-energy equivalency and, thus, will be affected by the pull of black holes. Consciousness has no physical constraints. It will be

waves that, as we all "know," usually propagate as fast as anything and always in a straight line! When it comes near a black hole, even its trajectory will be bent, and it will be sucked into the hole with no chance to ever escape.[115]

Dark matter is entirely inert, it has seemingly no potential of unfolding or bringing alive anything that can be conceived of as *physically* useful – but it does have undiminished gravity.

Since dark matter has no dynamism left and no longer has the capability to maintain *any* physical processes, the term "time," as we know it, has become meaningless with regard to dark matter.[116] Time is the elementary unit to measure *change*, i.e., the difference of *before* and *after* conditions in physical processes. In the absence of physical processes, every physical condition "before" is always the same as "after," regardless of how much time has elapsed in-between. This inactivates the laws of physics when it comes to dark matter (with the exception of the law of gravity, which appears to be the only mechanism by which dark matter can still exert influence in the physical reality).

– and remain – entirely unaffected by the brutal force of black holes. Even when all physical matter is burned out and converted to inert dark matter, the consciousness that has been "created" parallel to this life process of the physical reality will remain.

[115] The fact that light waves can be deflected by strong gravitational forces is actually physical proof – one of many – that light not only has wave but also matter/particle character. This wave-particle dualism has, of course, been one of the major facts of physics I have used in the hypothesis of the counterpart/ spiritual reality advanced in "*Consciousness or Entropy*?" It should also be pointed out that this only applies to "physical" light, described with DeBroglie wave lengths[3)] and propagation; it does not apply to spiritual "light" (see Chapter 7).

[116] Similarly, the concept of "space" has all but vanished.

Since most of the energy of matter has been converted to (unusable) entropy during the "deconstruction" process from matter into dark matter, and since entropy is an element that belongs in the counterpart/spiritual reality, as I have described earlier, dark matter should also be understood as belonging to the counterpart reality. This is actually a circumstance that attributes significance to dark matter, as we will discuss later in more detail.

The counterpart reality[117] is understood as the realm of the "implicate order," while the physical reality is the realm of the "explicate order." We conclude, therefore, that the process of burning-out of matter, i.e., its *transition* from "live" to "dark" matter, is nothing other than a return from the explicate to the implicate order.[118] It's the completion of a cycle that started with the Big Bang, creating explicit order ("live" matter) from implicit order (supreme-original energy) and now returns to implicit order.

For reasons of completeness, I remind here that the second half of this process (the returning to implicit order) is nothing other than the physical *energy* → *entropy* process. As we discussed earlier in this book, this process is paralleled with a dualism counterpart, the *thought* → *consciousness* process.[119] While energy, in the

[117] Or, more precisely, the *Supreme-Original Reality*, which contains the counterpart and physical realities as dualism.

[118] The choice of words in this sentence ("transition," "live," "dark," "explicate," "implicate") suggests the intent to include not only inanimate but also animate "matter" in this statement. There is an implicate suggestion that physical death of a person might entail a transition to a non-physical state, rather than a return to *nothingness*. If one combines this with the suggestion discussed in detail in Chapter 1.9 that there is a *process of becoming conscious (process of increasing consciousness)* that runs parallel with the process that ends in dark matter (i.e., the entropy process), one sees a powerful alternative to the depressing outlook many people have that there is nothing after death.

[119] See also Tab. 1.3 (Chapter 1.9) and Fig. 1.1 (Chapter 1.10).

process of turning into dark matter, has lost all its power, a new kind of power, consciousness, has emerged.[120]

[120] However, the *thought* → *consciousness* process becomes really significant only when physical beings capable of reflective thought are involved. We can probably assume that the *vast* majority of the conversion from "live" to "dark" matter occurred at a stage of evolution pertaining to that particular celestial body when no species capable of producing thought had emerged, and the overall yield of consciousness within this process of galactic dimensions may so far have been small.

3.6 No-Thing-Ness is Not Nothingness*

> ***Whoever it was who searched the heavens with a telescope and found no God would not have found the human mind if he had searched the brain with a microscope.***
>
> George Santayana

Considering that dark matter represents the majority of physical existence, it is justified to hypothesize what dark matter would be when viewed in the expanded frame. Is there *usefulness* that can be ascribed to dark matter? What would it be? Are there relationships involving dark matter?

In physiology, one is concerned with *interconnected relationship processes*. What kind of interconnected relationship processes could there be under *no space - no time* conditions? What would the *physiological interconnected relationship processes* be like in that case? Considering that, when viewed from the *physical* reality frame, there can be *no* physical processes at all in dark matter, can there be any *non-physical* (or "meta"-physical) processes? Given that a physical process is a *sequential* event, can there be *sequential* events in physical timelessness? Are these not mutually exclusive?

Dark matter is *no-thing-ness*, not *nothingness*!

As we have described, the "collapsed matter with no space or time contents" must be considered *structurally* defunct, if we apply the *physical* sense of the term "structural." *No (physical) things* exist, "*no-thing-ness*" reigns. The *physiology* of the Universe which we set out to hypothesize, develop, and analyze would then be the "interconnected relationship processes" of the *no-thing-ness* of the Universe. How can we approach and describe such a situation?

First of all, we must recognize that "*no things*" and "*no-thing-ness*" are not the same as "*nothing*" and "*nothingness.*" In fact, the term *nothing* typically only means that there is an absence of what we refer to as "things" in our *physical* viewing frame, i.e., the frame we are accustomed to using for discovering the physical laws of the Universe.

In our western society, we are entirely conditioned to relate worthiness to "things." Something that is not tangible *cannot* be worth anything! If you cannot touch it, weigh it, see it, smell it, perhaps hear it, or otherwise use it for physical processes, it is – by western definition – *nothing*.

Dark matter is not insignificant but may well be highly significant; it is what is on the other side of space and time.

In contrast, the "*no things*" in this context actually represent more than one-half of the physical existence (of the mass-energy equivalency of the Universe), and it would, thus, be entirely inappropriate to consider them not worth anything. In fact, they may more likely be the *main elements* ("significant-no-things") in the Universe.

Significant no-thing-ness is, thus, potentially the main outcome that has (so far) arisen from the event of the "Big Bang" event, i.e., we can argue that not "*nothing,*" but "*significant no-thing-ness*" resulted from it. Dark matter is not insignificant but highly significant; it is what is "on the other side" of space and time.

What would the elemental characteristics of no space and no time be? What would the "significant-no-thing-ness" be like? The answer, for me, lies in the hypothesis of a dualism counterpart reality, as I have described it in earlier chapters in this book. We recall that the counterpart reality can be described with the four physical (space-time) plus an additional dimension (the speed of light), whereby *processes* (i.e., sequential occurrences) do exist in the counterpart reality but occur so fast that, from the *physical* reality frame of reference, they occur concurrently and instantaneously.[121] "Mass" is an element of the physical reality. There is no mass, as we know it, in the counterpart reality, i.e., *no-thing-ness* reigns, but not "*nothingness*." The processes that do exist in the counterpart reality[122] refer to *no-things* but not to *nothing*.

If dark matter has significance, it must be postulated that there are relationships and processes between dark matter. What we conventionally call physical space and time can be excluded from any potential involvement, because they do not apply to dark matter. The only known common physical link between dark and "regular" matter is gravity.

To look at the importance of no-thing-ness in more detail, we must, therefore, first discuss the physical phenomenon of gravity in this context. We will do this in the following subchapter, in conjunction with the latest findings about neutrinos which literally "fill a gaping, *two-level* hole" in the understanding of the Universe and, as we will see, bridge between the two dualism realities (physical and counterpart realities).

---

[121] The reader may be reminded of the example of the "Super -Super Sonic Transport" given in Chapter 1.7.

[122] Thought - consciousness processes (i.e., processes of "becoming conscious").

3.7 Gravity and Neutrinos: A Bridge between the "Physical" and Nonphysical Realities?*

When you see a perfect 3-pointer basket thrown, perhaps the one that decided the game, you wouldn't likely be aware that it was what I consider the most miraculous law of the physical universe which gave this ball the parabolic trajectory that made it land just perfectly in the basket. Like invisible strings that stretch across the entire universe, gravity holds *everything* together. It is the weakest of all forces between small objects, but when they agglomerate to huge masses and attain intergalactic distances, the "gravitational micro-forces" add up to gigantic tensions that cause planets to circle around suns and galaxies to spiral.

Gravitational "field lines" exist *everywhere* and in *every possible direction*. Because they are not of an electromagnetic nature, they *permeate everything* in the entire Universe.

What else do we know about gravitational forces? We know that, in principle,[123] *each and every* atomic nucleus exerts a gravitational pull, ever so small but greater than zero, *on each and every other nucleus in the Universe, regardless where it is located.* It's like little strings (field lines) going out everywhere from each atom, interweaving with those from other atoms countless times, and thus creating a web so dense that these gravitational field lines

[123] In reality, of course, these gravity field lines are not separate entities and cannot be separated out individually. They constitute, instead, a "field" that is not dissimilar to the waters of the ocean that are composed of water molecules but are, for all *practical* purposes, one massive "body" of physical matter.

are *everywhere* and in *every possible direction.* Because they are not of an electromagnetic nature, they *permeate everything*, be it vacuum or dense matter, *as if they were entirely unaffected by anything that is of conventional physical nature.* So much for the "vacuum" that the engineers keep talking about[124] as being out there in space! We call it a vacuum because it – apparently – is devoid of matter. There are no physical particles.[125] But it is, in reality, *everything but an empty space* when it comes to the spheres of influence that are exerted between matter in space.

> ***Neutrino* are extremely fast moving tiny neutral particles carrying a minute, yet non-zero mass. They exist in an extremely high density and *permeate everything* in the entire Universe.**

It has recently been reported that tiny neutral particles, called *neutrinos*, carrying a minute yet non-zero mass, exist in an extremely high density[126] all-over the Universe and fill what we used to consider empty space. Similar to gravity field lines, they are electromagnetically neutral and, consequently, permeate *everything.* They zoom around at super high speed and are absolutely *everywhere*, including *inside* all physical matter, i.e., the rocks, the core of the earth, the oceans, the flowers, and you and me. The cir-

---

[124] This is by no means intended as a pun against the engineering professions. However, engineers usually operate more applied, more practically oriented than physicists, for whom the *vacuum of space* is actually a *pressure* (of approximately 10^{-10} Atmospheres), whereas the engineers would, more likely, look at space as an environment that has no impact on objects they might be designing for deployment in space.

[125] Actually even that is a gross misconception: even in the best vacuum, such as the vacuum prevailing in space, there are m*illions* of atomic species floating around and flying through each cubic centimeter of space each second.

[126] Billions of neutrinos per cubic centimeter exist anywhere in the Universe.

cumstance that they carry – albeit a very, very small amount of – mass conceptually links them to gravity.[127]

[127] In fact, it is believed that the combined mass of all neutrinos – even though extremely small *per individual* neutrino – adds up to such a huge amount that the majority of the mass of the universe heretofore unaccounted for and believed to be accumulated in *Black Holes* (some 99% of the mass of the Universe) is attributed to the neutrinos. Neutrinos are thus the most prolific *stuff* of the Universe.

3.8 Neutrinos and Gravity Fields as Dualism*

It is conceivable that *neutrinos* and *gravity field waves* form a dualism pair, i.e., that – similar to the photon/light-wave dualism of light – the dualism counterpart to neutrinos would be gravity field waves. (In this analogy, the neutrinos would be equivalent to *mass*, and the gravity field would be equivalent to *waves*).

***Neutrinos* and *gravity field waves* appear to form a dualism pair in the counterpart reality, similar to the photon/light-wave dualism that exists in the physical reality.**

There would, however, be a significant difference between neutrinos and conventional matter. The latter is essentially stationary or moving at relatively low speed, whereas neutrinos are already moving around *at very fast velocities* (it is assumed that they move at the speed of light). Consequently, it is necessary to assume, for that reason alone, that their wave equivalent (gravity field waves) propagate *at even much faster velocities*, which would mean that they clearly exceed the speed of light that had been believed to be the limitation of the physical reality. This would mean that gravity fields are phenomena that exist in the counterpart reality, i.e., they are not of a conventional physical nature (and cannot be measured with conventional physical equipment or means).

The assumption that gravity waves propagate at speeds many orders of magnitude faster than the speed of light *should be expected* for other reasons. Gravity is the only known "communication" mechanism between celestial bodies. It would be unthinkable that, for example, a comet (in the solar system) would not *instantaneously* change its course in the event of a major, sudden loss of mass, such as could conceivably be introduced by a nuclear explosion deliberately induced to change its course. If the gravity information about the comet's loss of mass would reach the sun not until a few hours later, the comet's course correction would also net get started until some hours after the explosion had occurred. Such an outcome would be absurd.

Gravity may actually be the *mechanistic link* between the physical and non-physical realities.

This conveys a picture of *total and complete interdependence* between everything that exists in the universe. The interdependence mechanism extends, via gravity, beyond the physical reality, and the effect of this interdependence is, therefore, expected to also extend to the non-physical realm.

In this sense, gravity may be more than an *example* in favor of the argument that a non-physical realm exists: it may actually be the *mechanistic link* between the physical and non-physical realities.[128] The continuum of gravity field waves permeating everything and propagating with essentially infinitely high velocity[129] might then well be the medium within which thought waves propagate.[130] They may well be what former generations of scien-

---

[128] These two realities are, of course, understood as being dualism counterparts of one all-encompassing entity, which we have called "Supreme-Original Reality" and, in later chapters, simply "God."

[129] I am tempted to speculate that they actually do propagate with the same general range of velocities we have earlier attributed to "thought waves," i.e., at somewhere around 10^{10} to 10^{30} times the speed of light.

[130] Gravity waves might, in fact – and I am going way out of line and out on a limb with this hypothesis – *be functionally identical* to thought waves. If that hypothesis were true, there would have to be a mechanism by which thought, such as produced in the mind of a person, can alter/affect gravity waves. This could possibly result from the physical/chemical/physiological change that takes place in the brain when/while thought is "produced." This change can be translated into an – ever so small – change of energy state of certain cell molecules,

tists were looking for when they were searching for the "ether" that would be the "medium" upon which light (and electromagnetic waves) would propagate. They may well be more, they may also be the medium upon which thought waves propagate and in which consciousness resides.

---

consuming – even though just a minute amount of – energy, concomitantly decreasing – every so slightly – the mass of the affected atoms and molecules, and consequently emitting slightly altered gravity waves. Since each atom is *constantly* in gravity wave communication with everything *everywhere*, this ever-so-small change is then emitted into all space and in all directions and can, thus, *in principle* be detected anywhere at essentially the very same instant it was emitted. The detection could have a physical, a quasi-physical, or any number of other conceivable manifestations. It could be "received," "stored," "reprocessed," or dealt with in any imaginable or yet unimaginable fashion. Here the limb becomes too thin for me – the reader may continue to "walk" further out on it if he/she is so inclined.

3.9 The Physical and the Non-Physical are One*

It has become apparent that I have been making a calculated distinction between the *physical* and *nonphysical* realms. I have labeled everything that is subject to velocities lower than the speed of light as physical; higher velocities characterize nonphysical phenomena. This distinction should be understood as specific to my writing; it is not intended to be re-introduced into science at large, for which there is only one reality, one that is (from the natural scientist's point of view) physical and has no boundaries. I am making this distinction for the one and only purpose of – paradoxically – emphasizing how weak the gap between science and spirituality has become. In reality (pun intended), *any* distinction between the physical and non-physical would be arbitrary, and there could only be one logical conclusion, which is that no boundary exists between the two but, instead, the two are one!

> ***Any* distinction between the physical and non-physical realms would, in principle be arbitrary. No boundary exists between the two but, instead, the two are one.**

It is remarkable that the physical sciences are, more and more, recognizing that the Einsteinian speed-of-light limitation no longer holds. Proof of the Bell theorem[131] in the early seventies, for example, clearly involved velocities greater than the speed of light and started this trend among scientists. Now that we *know*

---

[131] In 1964 J.S. Bell published his theorem. It was cast in terms of a hidden variables theory. Since then, evidence has been published by d'Espagnat, Stapp,[12)] and others that does not require hidden variables. See also Nick Herbert.[19)]

(not just *postulate* or *assume*) that a change of a physical characteristic (the spin) of one particle can be detected *instantaneously* by another, *regardless of the physical separation* (i.e., even over light years of distances), the magic of the barrier presented by the speed of light is gone – and with it the justification to separate the physical from the non-physical, i.e., what is "real" from what is "spiritual."

The knowledge we now have about neutrinos, gravity fields, and related long-range matter-matter interactions, makes the "detour" of distinguishing between the physical and non-physical realities *unnecessary* ...

The physical and non-physical realities do exist, but they are one and the same. The physical extends all the way throughout the spiritual, and the spiritual realm entirely pervades the physical reality. There is no difference; they are all one. The concept I presented, just a decade ago, by postulating the existence of the spiritual reality as a dualism counterpart to the physical reality – even though this concept still holds in all details then described – is no longer *required*. The knowledge we now have about neutrinos, gravity fields, and related long-range matter-matter interactions, makes this "detour" unnecessary.

... but I do feel that there is a redeeming quality in the dualism description of the physical and spiritual realities, and it is for this reason that I have continued using the concept in this book.

However, having made this important statement, I do feel that there is a redeeming quality in the dualism description of the physical and spiritual realities, and it is for this reason that I have opted to continue using the concept in this book. It allows, by drawing analogies, a degree of insight into the spiritual realm that would otherwise be difficult to deduce. It permits us to look at the situation from a familiar frame, the Newtonian-Einsteinian frame, which has given us a common sense of the dimension of time.

3.10 The Eternal Now*

In the beginning God … in the end God.

Desmond Tutu

We have now established that a higher-dimension (aspect of our) reality must exist, which can be conceived of as a dualism counterpart to what we used to call the physical reality. Whereas the physical reality is characterized by "*things,*" "*no-thing-ness*"[132] prevails in the counterpart reality, where *process velocities* are extremely high, in fact so high that, from our classical/Einsteinian physics frame of reference, they essentially occur *simultaneously*, i.e., timelessly.

What are the implications of timelessness? In the counterpart reality frame, occurrences that happened yesterday (or last year, or last millennium, or millions or billions of years ago) are perceived at essentially the same "time" as what is happening now *and/or what will be happening in the future.* The past, present, and future are one! The implications are enormous. If there were a mechanism by which conscious intervention from the counterpart reality in occurrences in the physical reality could actually happen,[133] such intervention would come from a frame of reference that is not bound to our physical time and space restrictions. Consequently, "they"[134] could *predict* future physical events or even

[132] No material "things" exist (see Chapter 3.6).

[133] We know from many records of "supernatural" occurrences that such interference *can* occur.

[134] This little word "they" – often used in a spiritual context – refers, of course, to conscious entities of any imaginable form or state that might exist in the non-

implement cause-and-effect interventions of galactic magnitude and over galactic distances.[135]

The implications of timelessness include that *nothing* is impossible.

What this really means, then, is that *nothing* is impossible. This is the reversal of an entire mind-set of conditioned attitudes that taught us, heretofore, to stick with what we can see, hear, touch, smell, calculate or, at least, rationalize and to beware of everything else, all those beliefs, feelings, notions (and often emotions) which one could label superstition and "should" consider unworthy of our human maturity. How enlightening! Talking with the dolphins is no longer such a far-out notion. Stepping on sacred soil becomes a worthy awareness. Beaming positive thoughts no longer is a futile activity! Wishing somebody well becomes more than just words! Saying "yes" becomes significant only when backed up with the *intent* of "yes" (and not "no"). All of these effects are timeless. They translate (essentially) instantaneously from location to location, and they remain active (essentially) indefinitely.

---

physical reality.

[135] Consequently, perceptions we used to call *superstitious* might not at all be entirely unrealistic when viewed in the enlarged frame we are talking about here.

3.11 The Universe as a Living Body

> ***Maybe the tragedy of the human race was that we had forgotten we were each Divine.***
>
> Shirley MacLaine

We can now continue with our attempt of looking at the Universe with a *physiology* based viewing frame. We can directly compare this macro-reality with our human societies, made up of our physiological human beings, and compare how all interact and interrelate with each other and with the atoms and cells that form them. Similarly, the Universe is a huge physiological Gaia system composed of "galactic societies" that can be likened to the human societies, each of which consists of uncountable multitudes of stars that are comparable to the cells in the human body. Commensurate with the size relationships between the macro reality and our terrestrial (micro) reality, long-range and short-range forces hold everything together and become the required communication links. The speed of light limits short-range forces, which are the predominant forces in the micro-world. Long-range forces[136] are predominant in macro/intergalactic frames and interact faster by many orders of magnitude.

Psychological and physiological interactions on societal and individual human levels appear to be unrestricted by speed, which indicates that they might occur predominantly on the basis of, or with means of, long-range forces.

Modern physiology, combined with other disciplines like psychology, medicine, alternative medicine, etc., teaches that in a

[136] Long-range forces are presumably mainly gravitational forces.

human body everything has its place and significance. There is no organ without a purpose. We used to think that certain parts of our bodies, such as the appendix or the tonsils, are dispensable and have no, or only negligible, impact on the well-being of the person. We now know better and no longer surgically remove these organs, unless there is an overriding, serious medical indication to do so. Similarly, we must conclude that *everything* in the Universe has its function and purpose. This includes the earth and everything on it – the dolphins and the squirrels, the redwood tree and the dandelion, and you and me. The thought that we are at liberty of destroying or tampering with our ecosystem, without affecting the larger context of which it is a part, becomes ludicrous.

Just as in a human body everything has its place and significance, we must conclude that *everything* in the Universe has its purpose. This includes the Earth and everything on it – including you and me.

The entire realm of health and health care could, in principle, be extrapolated to the galactic level. Skin cancer starts with as little as one cell that gets mutated, for example, by the impingement of an ultraviolet photon at the "wrong" spot at the wrong time. Its effect, if not properly treated, can spread to comparatively vast dimensions and eventually take the life of the person. Similarly, we must realize that we, as beings with the capacity of reflective thought, have the concomitant capacity to inflict "galactic cancer" to the entire system that we are part of. With thought comes responsibility. We do not live in a vacuum. What we do is not an isolated occurrence that has no further impact other than on ourselves!

And then there is the entire realm of feelings that can be translated to the Gaia concept. Can we imagine that a galaxy has feelings? That it can feel, happy, sad, inadequate, irritated, or whatever the feeling might be that we as human beings can associate with? Whatever it is that evokes feelings in humans, perhaps something we speak, even something we *think* about another person, could be translated *in principle* to the Gaia – galactic level.

Can we imagine that degree of influence we thus have on the Universe? If we would only understand a speck of that possibility, we would most certainly behave differently as human species. Love, compassion, consideration, tolerance, moderation, respect, would be much more common than they are now. Indeed, a loving smile can go a lot further than merely making another person happy!

CHAPTER 4

Expansion of Consciousness

Consciousness is a fundamental property of the source of all being ... more fundamental than energy or matter.
Elisabet Sahtouris

4.1 Intended Evolution?

In the earlier chapters we have discussed that the entire Existence can be understood as a dualism of a physical and a dualism *counterpart* reality, which we also earlier called *spiritual* or *gnostic* reality. The entire physical reality is subject to the energy-entropy law that describes a unidirectional process. At a certain point in the history of the Universe, the total amount of entropy was zero, and all energy existing at that time was "usable" physical energy. This event is generally called the "Big Bang" and happened some 13 billion years ago. As has been pointed out, a wealth of well-founded arguments exists that corroborate this theory.

The "Big Bang" was the result of *Supreme-Original Thought.* This supreme act of original creation prepared the way for evolution of the Universe and the Earth, and eventually of human consciousness to take place.

The Big Bang was the birthing point of time and space. Before, there was an *eternal presence*. There was no basis for physical processes to occur, and there was, consequently, no basis for physical evolution to occur. No mass or matter existed, and all energy was in the non-material state (*Supreme-Original Energy*).

In agreement with the dualism principle, the big, cataclysmic space-time birthing event would have had to be concomitant with, or preceded by, conversion of some of the *Supreme-Original Energy* into usable (physical) energy. This action must have been carefully reflected upon and executed, because it included the potential for all that would ever develop from it. It was a manifestation of *Supreme-Original Thought*.

The modern theory of evolution explains quite well what might have occurred after this original "creation" event that resulted in the first formation of the building blocks of physical matter. Voluminous literature exists in this field. The earlier referenced work by Pierre Teilhard de Chardin[36)] is one significant document on this subject. How the *details* of evolution actually unfolded from this point on is really of secondary significance, as far as the intent and direction of this book is concerned, since the entire miracle of creation was already completed at the very beginning, at the time of the Big Bang, where all theories of evolution begin. All the rest, howsoever intricate the development was, the sun, earth, water crystals, first living cells, fish, first mammals, and finally the emergence of man on Gaia Earth, all evolved in accordance with this "divine master plan."

The evolution of man with the capacity of creative thought may well have a major role in the entire underlying Divine Plan.

The evolution of man with the capacity of creative thought is no exception. As I discussed in "*Consciousness or Entropy?*" in some detail, the human may well have a major role in the entire underlying plan. One can even justifiably argue that the evolutionary process on our planet, which resulted in a maze of complexity, was in fact a *prerequisite* for attaining the intended outcome of the underlying plan (for our planet), which appears to include the development of high levels of consciousness. Without the human species and its capacity of reflective thought, little consciousness indeed would have had a chance to develop (within the realm of Gaia Earth).

4.2 The Evolution of Consciousness

I don't know Who – or What – put the question, I don't know when it was put. I don't even remember answering. But at some moment I did answer Yes to Someone – or Something – and from that hour I was certain that existence is meaningful and that, therefore, my life, in self-surrender, had a goal.

Dag Hammarskjöld

From the rationale developed in this book regarding the structure of the physical and spiritual realities, following a rigorous application of the dualism principle in physics, it must be assumed that the underlying plan includes goals that go beyond the creation of a physical universe. Instead, I would submit that the creation of the universe was only the *vehicle* with which the real Plan could be pursued. This plan can be hypothesized to be the attainment of the highest level of consciousness. This hypothesis has an interesting consequence: the universal intelligence (the "mind of God"), although existing from the very beginning of time, might then also in some way be evolving. Consciousness is increasing in time, and one instrument for this increase is the human species. One way to get from *Supreme-Original Energy* to consciousness is through a human being's mental action in the physical reality. The evolution has been unidirectional: *Supreme-Original Energy* was converted to physical energy, which in turn is being used to "produce" complexity and eventually consciousness.[137]

[137] See Chapter 1.10 for a pragmatic description of the *energy → consciousness process*.

The creation of the Universe is the vehicle with which the Divine Master Plan – the *attainment of the highest level of consciousness* – could be pursued.

In as much as the entropy resulting from any one physical activity does not disappear but "stockpiles," consciousness is not a momentary phenomenon that comes and goes. It is not affixed to a certain physical entity or process (such as life itself) and does not vanish when that physical entity ceases to exist (with death). Consciousness rather remains as some sort of collective phenomenon. As time goes on, the collective consciousness is increasing, i.e., is adding to the essence of the *Supreme-Original Entity* (God) and enhancing all that is and will remain.

As far as planet Earth is concerned, one can clearly argue that the underlying goal of enhancing consciousness might not have had a chance of success without the advent of man. The evolution leading to man must therefore have necessarily been the original intent and focal point of the underlying ("divine master-") plan.

The *Mind of God* is evolving itself – with the help of the human species!

It should be mentioned again at this point that one must be cautious with respect to over-zealously applying the dualism principle. The principle might suggest that some sort of proportionality exists between the amount of physical energy consumed and the amount of consciousness "produced." Such a notion is certainly an oversimplification. Whereby the formation of inert matter, like the planet Earth itself, took an enormous amount of energy (i.e., mass), the degree of complexity associated with inert matter is low. In fact, it might well be argued that no contribution to consciousness was made at all during the first few *billion* years following the coming into being of our planet. The human being, on the other hand, contains comparatively little physical energy (little mass), but his complexity is enormous, and his potential to produce thought, i.e., to process energy in the spiritual reality and to contribute to collective consciousness, is essentially unlimited.

The evolution leading to the human being as vehicle to add to collective consciousness, must have been the original intent and focal point of the underlying divine master plan.

I submit, however, that the mere production of thought is not equivalent with production of consciousness. Thought expended for materialistic purposes (i.e., to increase the complexity in the physical reality) is "knowledge," not consciousness. It takes thought expended with purely unselfish motivation to produce consciousness.[138]

[138] See Chapter 4.6 for a more explicit description of what I mean when I talk about consciousness.

4.3 Soul and Consciousness

You should not say, "God is in my heart," but rather, "I am in the heart of God."

Kahil Gibran

I have hypothesized that the soul is the dualism counterpart to a human being in the non-physical *counterpart reality*, which I have also called the *spiritual reality*. By analogy, the soul is then an entity that can function in the spiritual reality as freely as a person functions in the physical reality. Furthermore, in agreement with the dualism principle, the soul would be expected to have some (albeit limited) access to the physical reality. Just as human thought is limited in the spiritual reality in that he does not directly have the determination over the *outcome* of his thought "action," the action of the soul is limited in the physical reality in an equivalent way.

The soul is the dualism counterpart to a human being in the spiritual reality. A human being and his soul form a complete entity.[139]

[139] In approximately 1916, the now widely unknown Joseph S. Brenner published a book entitled *The Impersonal Life*. In this book, which reportedly came about in intuitive automatic writing, he presents, in remarkable detail – albeit in different language – a picture of the unity of a person's physical and spiritual self (his soul) that is remarkably similar to the person/soul dualism I am discussing here. Brenner's work had become the favorite book of none other than Elvis Presley, who – little known and publicized – was a devout spiritual person. Brenner's book was recently re-published by Jess Stern.[33)]

In accordance with the dualism principle, the soul is then vastly more significant than a human being, just as the spiritual reality is vastly more significant than the physical reality, since it has dominion over an additional dimension. The soul's functioning capability within the physical reality may, therefore, be perceived as less limited than a person's functioning ability in the spiritual reality. The soul, operating in the spiritual reality, may be able to oversee what is going on in the physical reality, but it may be impeded or incapable of intruding in physical actions.

The soul's functioning in the physical reality is similarly limited as a human being's functioning is in the spiritual reality.

In order to imagine more easily the consequences that dominion over an additional dimension might have, the reader may be reminded of the example of the impartial viewer of the ant moving in a two-dimensional plane.[140] The viewer, who has access over one more dimension, can perceive the ant in such a way that, in the frame of reference of the ant, he is not subject to time. He is able to see at one glance where the ant has been and where it is going. Similarly, the additional dimension in the spiritual reality may essentially eliminate, for a viewer "residing" in the spiritual reality, the effect and dependence of time. The soul may, therefore, be able to oversee the entire history, many generations, and the environment of the family of a person to be born. It may also have a very good oversight of what lies in the future for that family, just as an observer of the ant can see that the ant is unknowingly moving toward, and getting stuck at, a patch of glue in its path.

Birth and death are phenomena of the physical reality. They are interconnected with the *energy → entropy law* that is enacted in the physical reality and manifest in the spiritual reality.

---

[140] Chapter 1.5

The elimination of the physical time dependence in the spiritual reality (from the vantage point of the physical reality), and particularly the law of ever-increasing consciousness, suggests that the soul, operating in the spiritual reality, is immortal.

The law of increasing consciousness suggests that the soul is immortal.

The underlying Plan, according to which a human being and his soul and their ways of functioning were designed, appears to be so great that it seems impossible that anything more ingenious could ever have a chance of being developed. The mere fact that a person, having a soul that may be quite highly developed in the spiritual reality has virtually no direct benefit of this in the physical reality is superb.

The underlying Plan, according to which the human being and his soul and their ways of functioning were designed, appears so great that it seems impossible that anything more ingenious could ever have a chance of being developed. The mere fact that a person, having a soul that may be quite highly developed in the spiritual reality, has virtually no *direct* benefit of this in the physical reality is superb!

In as much as we have supposed that mankind is the vehicle to fulfill the underlying Divine Plan,[141] our physical life may be the vehicle for further development of our soul. Such development would be greatly impeded if the physical person were fully aware

---

[141] ... at least for our solar system.

of the entirety of the consciousness that his soul already has. The underlying Plan has, therefore, provided that the soul does not communicate directly, unequivocally with the person. In fact, most people are not even aware of their soul and would probably deny its existence altogether. The person's experiences are (or seem to be) unaffected by intervention[142] from the soul. On the other hand, the soul does appear to have some mysterious, indirect channels through which it can send messages to its human counterpart.[143] *Direct* intervention of the soul in its human counterpart's physical life, would it be possible, might actually *jeopardize* the impact of the decisions that each person makes and has to make. This would hold particularly for maximum impact decisions that determine the degree of development the soul will experience during the person's physical lifetime, decisions that mark essential leaps in consciousness. Such decisions must be taken with all senses, and in freedom of thought.

The Divine Master Plan appears to generally provide for the *greatest* potential for advancement of the soul *in the absence of direct intervention by the soul* in it's human counterpart's decisions.

[142] I am using the word "intervention" in this context as an entirely neutral word, with neither negative nor positive connotation. It simply means: *attempt to affect a physical change; an effort to change the course or direction of someone's life*.

[143] It makes sense to consider dreams as such channels for messages. However, decoding requires maturity, self-knowledge, and an open mind from the person. Misinterpretations occur often, and the consequences of misinterpreted dreams may be so adverse that it may be better, in the absence of mature self-knowledge, to disregard them altogether. Other powerful means or methods include meditation and prayer.

4.4 The Longevity of Consciousness

Consciousness, once produced, will continue to exist *indefinitely*. It will not disappear, it cannot die; it is independent of the physical body that is subject to the life-death cycle. Not only will consciousness, therefore, not disappear with the death of the person, but it is "set into the atmosphere" the very moment it is produced. Where it goes, what it will do or undo, is out of human control.

Consciousness will continue to exist *indefinitely*. It will not disappear; it cannot die. It will continue to be available for "pick-up" anywhere in the universe at any time.

The following personal experiences may serve as an example for this important statement about the longevity and influence of consciousness. Over the past decades, I worked together with many people and observed many people and groups who were donating their time and effort to the general cause of betterment of humankind and our planet at large. I want to elaborate on two such "projects" as typical examples.

In 1974, a group of very motivated and dedicated people in the San Francisco Bay Area became concerned about the danger associated with the use and proliferation of nuclear power. They informed themselves about the issue and went out to pass their findings on to as many people as they could reach. Their methods included presentations to large audiences, gatherings in living rooms to discuss the topic, teaching college and high school courses on the danger of nuclear power, and prayer and meditation. In 1975, the work of these people culminated in a vigorous political campaign to get the *Nuclear Safeguards Initiative* on the California election ballot. Once half a million signatures had been collected, the proposition qualified for vote by the electorate. The proposition ended in a demoralizing defeat by a 3:1 margin. Many of the volunteers, who had worked so hard toward qualifying for the ballot and to curtail the spread of nuclear power in California, despaired. They considered their efforts failed.

But had they really? In 20/20 hindsight, 1975 marked the beginning of a world-wide change of course in the nuclear power industry. I remember that in some of our presentations we used "dot chart" maps of the U.S. where each dot represented one nuclear power plant. Three maps were compared: the then present situation (1974) and projections for the years 1985 and 2000. The 1974 map had just a few dots, about 50 in total for the entire country. Many states did not have any dot at all. For 1985, many more dots had been added, and for the year 2000 so many nuclear power plants had been projected by the electric power industry that the entire map was so densely dotted that it appeared dark (see Fig.4.1).

Everybody knows what really happened. Since the mid seventies, just a very few nuclear power plants, those that had already been under construction at that time, were completed and came on line. The total number for the year 2000 ended up actually at *fewer nuclear power plants than were already in use 30 years ago*. Few, if any, new plants had been built, and those that had been decommissioned due to reaching their service life were not replaced. This trend is not only valid for the U.S. but holds for the entire world. For all practical purposes, nuclear power has ceased to be a viable option for electric power generation anywhere in the world. Even in Germany, and now also in France, two countries that had relied exceedingly heavily on nuclear power,[144]

[144] In 1975, France derived more than one third of its electric energy from nuclear power plants. This number, albeit one of the highest in the world, has since declined.

decommissioned nuclear power plants are being replaced with conventional plants.

How did this happen? What was it *really* that had made the difference? Clearly, the defeat in the election did not matter. The outcome was even better for the proponents of the initiative than what had been projected in the most favorable winning scenarios. It was even better than in the optimistic dreams of any of us. In fact, it was believed that, had the nuclear safeguards initiative succeeded in the election, it would probably have been challenged in the courts and, likely, been ruled unconstitutional.

It would be highly presumptuous to think that these few hundred people alone could have had an effect as profound as this change of the way humanity would think about electric power generation. But there is also no doubt that this example does point to a cause-and-effect relationship that can, ultimately, only be explained with the notion that consciousness is timeless and works in uncontrollable, often mysterious, very effective ways. What started back in 1974 in the minds of a few people, quickly found early supporters who took it and put tremendous personal engagement into it. It then snowballed, even beyond our national boundaries, and became a world-wide movement that would pursue the idea, totally in its own right – and change the course of history. By 1976, for example, the idea had taken hold in Germany, where it quickly reached dimensions that surpassed the intensity ever attained in the U.S.

There is no doubt that this example does point to a cause-and-effect relationship that can, ultimately, only be explained with the notion that consciousness is timeless and works in uncontrollable, often mysterious, *very effective* ways.

Ever since that time, to complement the example, the consciousness about the risks of nuclear electric power generation has prevailed. When new challenges arose, such as the concern about global warming,[145] consciousness about the risks of nuclear power would continue to play a dominant role in the minds of people and governments. Instead, attention has consistently been directed at other means of environmentally benign production of electric energy, such as hydrogen fuel and its production with solar energy.

The other example goes back to the more recent historic effort to curtail the East-West arms race. It was pursued with the same kind of personal commitment and vigor than the education campaign against nuclear power generation. It started in the U.S. and in Europe independently and very differently. Around 1980, a comparatively small group of dedicated people began to raise the consciousness about the ineffectiveness of war as means to resolve international conflicts in general, and about the immorality of nuclear war and the preparation for it in particular. Within just a few years, the totally unexpected democratization of the East Block countries, and the dismantling of the East/West military confrontation, would ensue. So fast would this change of heart take place in these nations that not the rate of reduction of armaments by mutual arms reduction agreements, but the rate of *actual destruction* of obsolete weapons systems would become the managerial focal points. For example, the deactivation and destruction of the stockpiles of chemical weapons stored in the Taunus Region in Germany and of the masses of Soviet weaponry that became obsolete with the unification of East and West Germany in 1989, lagged, at times, months behind schedule. This followed decades of laborious attempts of negotiating *any* kind of weapons reductions, with exceedingly sluggish results at best. The power of consciousness can, indeed, be beyond any rational expectation.

---

[145] The greenhouse effect is "fueled" by natural gas and fossil fuel-burning power plants and would thus favor nuclear power plants that do not produce the detrimental quantities of CO^2 causing the global warming problem.

Once consciousness has been "born," it will take on its own life. It is no longer tied to the person who gave birth to it. I recently experienced a vivid example of this rule. Throughout their efforts to diffuse the nuclear arms race, the people working on this issue kept using an insight attributed to one of our astronauts traveling to the moon. It was a quotation that perfectly suited the purpose of bringing to consciousness that our planet is small and finite and, when viewed from outer space, is one intricate, interdependent system and knows no boundaries. In a beautiful way, the astronaut raises to consciousness how mistaken, how *absurd* it is to think that problems could ever be solved by dropping nuclear weapons on other people. Many years later, I had the privilege of personally meeting the author of the quote. Happy to finally be able to meet the famous astronaut whose saying had meant so much to me and many others over so many years, I asked him to elaborate on the insight he must have had. To my great surprise, his answer entirely centered around the physical situation that had led to the quote, problems with operating the camera when they tried to photograph the earth when they were emerging from the far side of the moon, and alike. It felt as if the author of the quote and the life that the quote had taken on had become disconnected. I also realized that this is not detrimental but, in fact, quite "human." The importance of the quote, or of any conscious utterance by any person, is not tied to the subsequent direction of the life of its author. Consciousness speaks for itself.

Once consciousness has been "born," it will take on its own life. It is no longer tied to the person who gave birth to it.

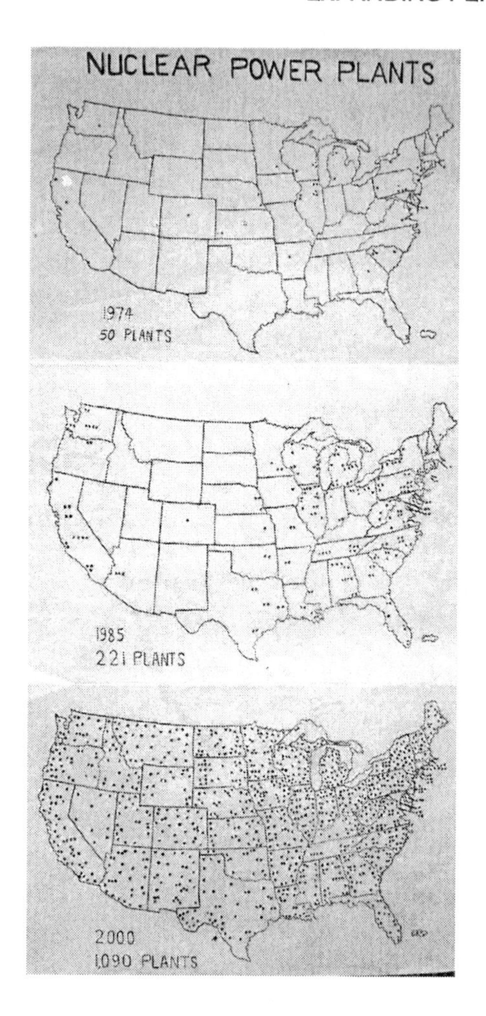

Fig. 4.1 Instructional charts used in the 1975 *California Nuclear Safeguards Initiative* campaign. Each dot represents one nuclear power plant, as projected in 1974 by electric power companies. The *actual* number of nuclear power plants in 2000 was, however, fewer than 50, i.e., *less than* what *existed* in 1974. The California ballot measure had "lost," but the consciousness raised during the campaign provided the desired results anyway, way beyond expectations.

Fig. 4.2 Women of various ethnic origins peacefully demonstrate (1975) in the streets of San Francisco for a world without nuclear power plants (in front/center: my wife Adelgund Heinemann).

Fig. 4.3 Even though the 1975 California *Nuclear Safeguard Initiative* was soundly defeated by the California electorate, its message was highly effective all over the world. Rather than a 20-fold increase of nuclear power plants, as was predicted by the nuclear power industry for the U.S. by the year 2000, we actually experienced a net *decrease* (stemming from decommissioned plants that were not replaced).

4.5 A Rationale for Reincarnation

Since a human being is "unconscious" at birth but has the birthright of becoming conscious during his life, it may well be that at the end of a person's life his soul may find itself "unhappy" with the degree of consciousness it has "attained" during its "excursion into the physical reality."[146] There may have been a special kind of consciousness it might have wanted to attain during this lifetime. Such a goal does not have to be some grandiose scientific advance. An unknown person who has found his or her inner center and has implemented personal insight, however minute it might appear to an outsider, in his or her own life may well be more conscious than a famous scientist who has never mastered the feeling of inadequacy or envy when comparing himself to his peers.

The soul's goal during a person's lifetime does not have to be some grandiose scientific advance. A person who finds her inner center and implements personal insight, however minute, in her own life may well be more conscious than a famous scientist or glamorous actress.

The special experience the soul may need to pursue in a physical lifetime, the "karmic knot" it might have to untangle, might be to cope with illness, or with wealth, or perhaps with some personal weakness. As a consequence, the soul might, following such "unsuccessful" incarnation, decide to incarnate again for an-

[146] See Chapter 7.2 for an expansion of this metaphor.

other try at solving its task. It is therefore not unreasonable to rationalize the occurrence of some sort of reincarnation.[147]

It would be interesting to examine the dualism principle and the analogies presented in this book with regard to the "selection process" that the soul might use, or be capable of using, when it "decides" to incarnate. Having consciousness and existing in the spiritual reality would imply that the soul has intelligence. In fact, its intelligence is probably overwhelming, because it is likely that the soul can communicate with other souls and can, thus, tap into the tremendous wealth of all past and present consciousness that has collectively been attained by mankind. Furthermore, it is likely that this consciousness is uncluttered from all the distractions that we humans are constantly subjected to. The soul has spatial and temporal omnipresence. Therefore, it will be able to evaluate a wealth of data regarding, for instance, the history of a family it is considering to be born into, its environment, its ambitions, its generic pre-dispositions, and alike. It will thus be able to make a very good "guess" as to the chances that the specific tasks it wishes to resolve will present themselves during the lifetime of the person it is about to become.

If the soul's "blueprint"[148] for it's life-time journey is not met during an incarnation, chances are it will need another incarnation to complete the mission.

---

[147] In Chapter 7.2, I am using a "study trip" as a metaphor for a physical lifetime of a soul. Just as we would make journeys with various objectives during our professional lives, it is conceivable that the soul undertakes various incarnations with different themes.

[148] The use of the word "blueprint" in the context of a person's life destiny was suggested by the renowned intuitive and author, Sylvia Browne[4)].

It may be helpful to point out that, in accordance with the dualism principle, as I have described it in this book, the spiritual reality is *not entirely* independent of time. It would only appear that way from our frame of reference in the physical reality. Processes occur in the spiritual reality with velocities many orders of magnitude faster than the speed of light. Nevertheless, processes do occur, even there, in a sequence, separated in time. It is, therefore, logical to assume that the soul, or any other being that exists in the spiritual reality, will not be able to actually *see into* the future. It can only *draw conclusions* about the future, how it will *likely* unravel in the physical reality, based on all data available from the entire past, but there is no *certainty* that the conclusions drawn on the basis of these data will be correct.

One might also think of the consequences that *upsetting* such a predisposition might conceivably have. For example, adoption into a different family, or simply divorce and re-marriage of the parents, might be an upsetting event. It appears that simply reflecting upon one's own life, the decisions I made that brought me where I now am, might be a very meaningful task. What did I do in my life that was truly unpredictable, what was out of the line of the "ordinary," what did I do that a super-intelligent being that was able to overlook the entire history of my family and my environment could not have "predicted?" And if there are certain unpredictable moves that I have undertaken, what predicament did they lead out of or into? Have they deprived my soul from even encountering those karmic knots that it wanted to untangle in my lifetime?

4.6 Love: the Highest Form of Consciousness

I believe in God, who is for me spirit, love, the principle of all things ... I believe that the reason for life is for each of us simply to grow in love. I believe that this growth in love will contribute more than any other force to establish the Kingdom of God on earth.

Leo Tolstoy

Among all theorizing and conjecturing, it often strikes me how powerful the truth of this quote from the famous Russian author really is: the road to consciousness is through love. Without love, what we might think consciousness entails is really nothing but information and/or knowledge.

This came home to me during this writing, when we learned about the broken hearts of two people who are both dear to us, following the break-up of their long-term relationship. Learning how to work through problems that involve the heart of other people is the battleground for our souls. We win when we lose. Superficial victory will bring momentary gratification. It likely never leaves the realm of the physical. But "real" victory, obtained in an outpouring of genuine love, will give birth to consciousness. Love is more powerful than anything we can imagine.

Whatever I could say here about love would be dwarfed by the teachings on love in the sacred writings and by the wisdom teachers throughout the ages. The following poem by *Emmet Fox* summarizes the essence of true love quite convincingly:

Love is by far the most important thing of all.
It is the Golden Gate of Paradise.
Pray for the understanding of love, and meditate upon it daily.
It cuts out fear.
It is the fulfilling of the Law.
It covers multitudes of sins.
Love is absolutely invincible.

There is no difficulty that enough love will not conquer;
no disease that enough love will not heal;
no door that enough love will not open;
no gulf that enough love will not bridge;
no wall that enough love will not throw down;
no sin that enough love will not redeem.

It makes no difference how deeply seated may be the trouble,
how hopeless the outlook,
how muddled the tangle,
how great the mistake;
a sufficient realization of love will resolve it all.
If only you could love enough, you would be the happiest
and most powerful being in the whole world.

As beautiful as this poem is, it – in and by itself – is not consciousness. It is a beautiful string of words. It is powerful and inspiring. But it is not consciousness. It is you, the reader, who can turn it into consciousness. If you are *touched* by it, if your eyes fill with tears when you read it and you don't know why, if you pick up the phone and call your mother and say to her, "I love you, mom!" – and ask yourself, "How come I did that?" – then you may be on the way toward becoming conscious.

That's what it's all about! Until you have taken it to your heart, the most beautiful poem, the wisest statement, or the most intelligent scientific finding is nothing but a piece of knowledge. Just as a key for patentability of an invention is its *reduction to practice*, the key for attaining consciousness is *reduction to the level of the heart*.

Love does not require a Ph.D. It is not found in books, nor in stores, nor in memories. It grows in your heart – not in that organ in your chest that pumps your blood through your veins and arteries, but in your innermost being, your true self.

CHAPTER 5

Expansion of Belief Systems

In the realm of the mind, what you believe to be true tends to be true -- so watch your beliefs.

Charles Tart (after John Lilly)

5.1 Synchronicity

Recently, when I was thinking and writing about if God can intervene *directly* in physical occurrences, I had a sequence of experiences that literally gave me the answer – and it was *not* the answer I had expected. It started within minutes after I had finished a paragraph describing my conviction that God acts *only* through human beings. It was early on a Monday morning. We were spending a mini-vacation at our favorite place on the California coast. Gundi had just emerged from our bedroom and was preparing our morning coffee, when it came into her mind that it would be nice to hear some music, such as one of the beautiful songs by Charlotte Church. While she was walking toward the stereo unit, Charlotte's beautiful voice started singing. I was sitting at my desk on the mezzanine and commented on the nice music she had just turned on. "I didn't turn on any music!" she replied, "I was just going to do that and then it started playing. Didn't *you* turn it on?" We had been alone in the house for the entire weekend, and neither of us has any idea whatsoever how the CD could have started playing.

I was still puzzled about this strange incident when, just a few minutes later, my secretary called and told me that our company had been awarded the big government contract for which we had submitted a competitive bid months ago but had assigned a low probability of success. The word was that the award had actually been made several days ago, that all unsuccessful bidders had already been notified, and that they had simply "forgotten" to pass the good news on to us earlier ...

The third unusual experience happened at the end of the day, after our return to our home in Sunnyvale ... the news about the contract had cut our vacation short, as it required my presence in the office. Before our departure a few days earlier, my wallet, with driver's license, cash, several credit cards, and a number of IDs that make you a person in our society, had disappeared and was nowhere to be found. I was able to reduce the window of time when it must have been lost to literally a few minutes, between visiting the ATM at my bank and welcoming dinner guests at our home the night before. I had checked and double-checked *everywhere*, including literally a dozen times in the pockets of the pants I had been wearing that evening, and I had eventually given up the search for the wallet and canceled all credit cards it contained, before we left for Sea Ranch. Now back at home, I checked one more time in my wardrobe ... and there it was, in the pocket of the pants I had been wearing that evening....

Coincidences? For *one* of the occurrences, may be. But for *two*, and even *three*? Of course, a logical explanation that could satisfy a rational mind could likely be found for each occurrence. In fact, it had to be found, because all occurrences in our physical reality do eventually get manifested by physical means. The CD player must obviously – albeit for reasons unknown to me to this day – have been programmed to start playing *exactly* at that particular time (!), and the CD magazine of our stereo contained an album with Charlotte Church's songs. And, *of course*, the wallet must always have been there, in that pocket, all the time, I *must* have simply overlooked it ... But I invite you be the judge about the *synchronicity* of these events – I have drawn my own conclusion. I had been taught a lesson. My belief system had been too narrow.

There is no doubt that, with the framework of the high-speed counterpart reality, phenomena of synchronicity can be *explained*. They are mind-to-mind "transmissions" that do not require physical means of "transportation," such as speech, radio waves, or alike. They can bridge over large physical spaces. There should be no doubt – even in the minds of critics – that such phenomena exist. They have been reported many times with a high degree of credence. Thought projections are not subject to the speed-of-light limitation but can occur at much faster velocities, in essence so fast that they appear instantaneous. The overall conclusion of what I had experienced that day, then, is that, while the occurrences can be explained with physical means, they make no sense to me in the absence of the a higher consciousness that had "interfered."

It is likely that synchronic events occurring in our lives *are* signals of a deeper meaning.

For those to whom the spiritual reality is not a doubtful concept but rather part of their belief system, it is natural to live with the notion that each of us has *spirit guides* who are around us much of the time and are available to assist us whenever we call upon them. Often they will try to communicate some important message to us, and one avenue of getting our attention is through occurrences that appear "unnatural" to us. I am sure that the above-described string of coincidental occurrences was an attempt of one of my spirit guides to tell me that I was wrong with my statement that God cannot interfere with physical occurrences. "Of course, He can," I was being told. "There is nothing the Spirit can *not* do! Even the most "impossible" task is possible to the Spirit – perhaps not very *likely*, but *not impossible*."

5.2 Premonitions

To me, faith is not just a noun but also a verb.

Jimmy Carter

Another way our spirit guides try to get a hold of our attentiveness is through dreams. I am sure we all have ample examples of how this works. One morning, at the breakfast table, Gundi shared with me a dream she had just "lived" through. The dream was about John, a good friend of ours. Thea, his wife, had called her to chat about John and his latest craze. "Guess what he is doing: he is chanting and drumming! He is totally convinced that this is good for his health!"

All who know John would very much question whether this brilliant engineer, who – albeit spiritual – would never touch anything that is not solidly anchored in the laws of our physical reality, would give in to such "weird" notions as drumming and chanting being good for his health. He would be the type who would undergo the most modern medical treatment if there were a medical indication, but chanting and drumming … no!

Two weeks after that incident, Gundi did get an out-of-the blue phone call from Thea. They had not talked for a while, and Thea wanted to let us know that John had come down with cancer. He had opted against surgical treatment and for radiation therapy. While he did not go quite as far as taking up drumming, he had adopted a strict diet combined with homeopathic medicines and was quite confident that this would lead him past his physical predicament.

A friend of our family, Mary M., reports the following dream involving her son Josh and his wife Anne. "*They had just moved into their beautiful new home in the hills of the San Fernando Valley, with breathtaking vistas – the kind of home John (her husband) and I had dreamed of living in ourselves for many years but could never quite afford. In the dream, Anne, with whom we have had an excellent relationship, called me to let me know that we would soon be able to move into their house. 'We won't need it any more!' Just a few days later, a terrible accident happened to Josh. Being an avid rock climbing expert, with numerous extreme climbs to his fame all over the world, he had gotten in trouble and literally fell some 200 feet (60 meters) straight to the ground. In what could only be described as a miraculous chain of events, it turned out, however, that his severe internal wounds were associated with only minor bleeding, rather than the massive bleeding that would normally be expected for such organ damage. And every single one of the many bone fractures he had suffered, starting with a C2 neck fracture, turned out to be in the relatively most benign positions. Within a couple of weeks, Josh was up and walking again, and a few weeks later he had fully recovered from an accident that, for all human rationale, would have been expected to be fatal.*"

Both stories are premonitions. However, their recognition as such at the onset is not straight-forward. It is much easier to draw that conclusion in 20/20 hindsight. That John's unusual new activity would announce his diagnosis of cancer could not be inferred from the dream. However, Gundi, who is a healing arts practitioner and is very much in *tune* with the healing effects of sounds and chanting, *did* draw the conclusion from the dream that there might be a likely connection to an illness that was in need of healing. And there is little doubt about the character of a premonition in the second story, but it was impossible to draw concrete conclusions at the time.

The striking facet of these stories is not only the accuracy of the premonitions Gundi and Mary had received in their dreams.

It is also the totally unexpected turns of events of these cases. But were they really *unexpected*? Not so in my mind! Quite some time ago, Gundi, and also Mary, had adopted the daily practice of beaming healing thoughts toward persons who they knew were in distress. As a matter of course, they had included John and Thea, as well as Josh and Anne, respectively, in their focusing activities, as soon as they had experienced the dreams, more so than they would have normally done.

Mary's account about Josh's accident would not be complete without mentioning yet another wrinkle of what had happened. "*Josh's friend, who was with him on this climbing adventure, had a cell phone and was able to call us and reach 911 within minutes of the accident. All he was able to communicate to us was, though, that there had been a severe accident involving Josh, and that he was alive. Upon this message, calm and confident, I started to focus healing energy toward Josh and literally activated a "prayer chain" of like-minded friends. For absolutely no overt reason – we had no information whatsoever about the nature of the accident or of the injuries involved – I intensified my focus of healing intentions on the very subject of minimized internal bleeding. To this day I see no* <u>*rational*</u> *reason for having done so. Internal bleeding had certainly not been a subject I had recently dealt with or that had any reason whatsoever to be in my consciousness. In fact, my general knowledge of trauma risks was sketchy enough not to identify internal bleeding as particularly high on the list of worrisome medical indications.*"

How did it come about that this subject became a prime focus for Mary at that very moment? The reader may draw his or her own conclusion. Mine is that premonitions, coming to us in dreams or other ways of intuitive perception, are real communications to us from what I have called the counterpart reality (the spiritual reality, the divine, the Gnostic, or whatever word you may use). I also conclude that, in as much as we can perceive such premonitions, we can turn the course of events by beaming positive thoughts in the direction of the concern thus revealed to us.

There may also be a simple, two-word description of the same conclusion: *prayer works*!

Focused attention can change a physical course of events. Premonitions occur to make us aware that our focused attention is of need. We can then turn the course of events by beaming positive thoughts in the direction of the concern thus revealed to us. There is no need to know details!

My own conclusion goes beyond that. The "communication" from the divine reality had a very specific character. It was personalized. It was tailored for the recipient. It was not just a symbol or a symbolic story. It occurred in a *very specific* way that meant something to Gundi and Mary but would likely not have meant the same thing to someone else having had the same dream.

Nothing involving human beings is fully predetermined.

The possibility of turning the course of events around by focused attention and intention is *extremely significant*. Of course, it underlines that focused intention does help. It also states that nothing involving human beings is fully pre-determined. Yes, it is true that certain events will take their course, but with the paramount qualifier, "*unless intentionally changed*." We can perfectly well predict where an airplane will end up when it is flying on autopilot, even if the underlying autopilot instructions are very intricate and include many course changes. We can even predict, in surprising detail if needed, what will happen when there is a certain malfunction of the autopilot system. But an "unforeseen," unpredicted human course adjustment/intervention *cannot* be predicted. Our endowed capacity of reflective thought allows us to decide for or against any relevant cause. We can decide for true love, or we can decide to remain indifferent. The latter allows predetermination to take its course. The former can change the course of history. We are the tailors of our history. No less, and no more.

We can decide for true love, or we can decide to remain indifferent. The former can change the course of history; the latter allows predetermination to take its course. We are the tailors of our history. No less, and no more.

Premonitions can also come in broad daylight. On one quite ordinary day, when I was opening an ordinary stack of personal mail, one particular letter caught my attention. It was a request for a donation – quite ordinary, not really too different from the countless requests for donation I keep receiving. But for some reason, this request did not get the same treatment as it would have normally gotten. It did not go straight into the wastebasket. Instead, I started to read it. It reported about a young man who had experienced severe injuries in an accident. He was paralyzed from the neck on down and needed money to pay for medical bills and the necessary lifestyle changes that would follow this tragic accident. Without a hesitation, as if on auto-pilot, I picked up the phone and called the number of the church that had sent the letter. The accident, I found out, was that the young man had been run over by a truck while riding his bike. My donation check was in the mailbox within minutes. In fact, I don't think I had ever before written a donation check in such sizeable amount that quickly, and with such an absolute surety that it was the right thing to do. With a remarkably unusual speed, the case was "settled" and I was back busy with my regular office work.

Later that day, my wife got a phone call from our daughter. Hardly able to talk, still somewhat in shock, she reported what had happened to her about an hour earlier. On her way home from a college class, she had been riding her bike straight through a city intersection, when a big 18-wheeler to her left, unannounced and unexpectedly, made a right turn and almost rolled over her. She was barely able to push herself away from the big rear wheel of the truck, missing being run over by inches. The truck driver never even realized what had happened and kept on going.

Putting two and two together was not only a lesson in premonitions for me, it was also a stark a reminder that our guardian angels are not resting idle....

5.3 No Thought is Ever Lost

Given what we now know about the spiritual reality, it is quite clear that there is no such thing as "losing" something. Nothing will ever get lost. No thought, no idea, no expression, no feeling, no twinkle with the eye. In a continuum where no time and no spatial boundaries exist, everything that falls in the realm of what we conventionally call "non-physical" has no chance of ever escaping. It will not just become part of an endless pool of indistinguishable entropy, total chaos, but will retain its original "form" in all detail.

There will thus be no need to frantically "finish" something, for the only sake of it not getting lost. How often are we preoccupied with the urge to complete an unfinished project? When the time is right, completion of the project will become a matter of course, almost like automatic. The thoughts that went into the subject of this book but did not yet get written down, or were not included for any reason, did not vanish into nothingness but, having been no-*thing*-ness from the very beginning, will be available for continuation and, perhaps at some time, by someone, at some place, be picked up and further pursued.

This may well be the mechanism by which knowledge has been perpetuated over the ages. Have you ever wondered why it is that little kids use a computer almost instinctively, like a toy, completely self-evident, as if it was the most congenial toy to play with? Was it like that when you started using a computer? If you are like I was, the experience was quite different. You had to *learn* it. There was *nothing* natural about it. In fact, you resisted it at first. When the first primitive PCs came on the market, I was a professor at Stanford University. One day, our Department Head announced that we could purchase a "personal computer" at a highly discounted, university-subsidized price if we would submit ourselves to a "test" after a few weeks to demonstrate that we could actually make use of it. Granted, learning the prehistoric "basic" computer language, without a mouse, pull-down menus and alike, was a far cry from the situation our grandkids are finding today, but it nevertheless seems like there is something "in the air" today, if not in their genes, that makes it natural for our grade school youngsters to pick up how to use a computer, with a snap of a finger.

It has been said that the amount of basic knowledge our university students have to cope with doubles every five years. What Einstein found in the physics books to study in his time is dwarfed by what our high school seniors must learn to pass an honors class in physics. How can this keep going on – in the absence of some sort of "automatic aggregate knowledge pickup" mechanism?

In accordance with the dualism principle, as I have pointed out earlier, the human intellect, and finally human consciousness, functions in the space/time-independent nonphysical ("hyperphysical") reality. It is, therefore, entirely feasible that we have some sort of receptors – of which we are entirely unaware – that enable us to pick up certain *essence* from that reality and add it to the data/knowledge base in our individual human minds. It is as if it has been added to our "basic software package," if we stay with computers as a metaphor.

When I purchased my first computer in 1979, it came with a basic *disc operating system* (DOS) and essentially nothing else. My TRS-80 was only useful to me because someone else had developed a "basic" computing language which, once introduced into the computer, enabled me to apply it to perform certain calculations I needed to get done, faster than with other methods then available to me. There was also the added attraction that I could use my new computer for writing. It was *far* superior to my typewriter, in that I could delete and correct, even insert new text at any place and thus much more easily edit a manuscript than I could ever have done before. I could even store my manuscript on a flat disk, "only" 5 inches in diameter, and print my text out at any time later, as often as I wanted. My computer, I had learned, had 16,000 bytes of memory, and at an extra charge of "only" the equivalent of a one-month mortgage payment on my house I could upgrade that memory to 48,000 bytes – more than I would likely *ever* use! The "small" storage disk would be able to hold as many as 30 pages of my manuscript – what a terrific improvement over having to carry thirty typewritten pages in my briefcase!

The laptop I purchased a few months ago, less than a couple of decades and a half later, came with a memory that was not less than 30,000 *times greater*! It can store *one Million times* more data (or manuscript pages) than my TSR-80 was able to. And my laptop came with a "software package" that lets me do things nobody had even heard or dreamed of at the TSR-80 times, such as connect to the *Internet* and instantly get information that formerly would only have been available from the most sophisticated libraries. It would not only let me write and edit my manuscript, it would also *automatically* correct my typos – often even without my noticing it – and allow me to add illustrations, *even color photographs*, to my texts. It contains the possibilities to send my text to my friend at the other side of the world, and he would almost *instantaneously* receive it and be able to read it and make editing comments to it. He would then be able to immediately send it back to me, and I could choose to accept or reject his editing remarks with a stroke on my keyboard.

All this additional information/data capacity comes, if you will, at no extra charge when compared to the TSR-80 times. All this came for less than a typical current monthly mortgage payment! It is just there, ready for use. Yes, it is now being used by millions of people. It has become part of the "givens" in our early 21st century lives. In fact, it has increased the speed with which we "produce" and "process" knowledge at such a pace that simply *not* using it would place us in an economically disadvantaged position.

Obviously, this kind of technological advance has been made available to us in many areas affecting our lives, and it is not unthinkable, or unreasonable, to assume that the same principle holds in non-technical areas, as well. Even though not quantifiably, but nevertheless in a most real sense, we benefit from the love and concern of others, from the examples set by Martin Luther and Martin Luther King, from the compassion of Mahadma Gandhi and Mother Theresa, from the superb intellect of Albert Einstein or Stephen Hawking[17)], and from the loving care of our mother.

5.4 Jesus as Teacher and Jesus as a Mystic: a Dualism

The study of Jesus of Nazareth and his teachings has had an important role in my life. In my childhood and early adolescence, commensurate with my fundamentalist Christian upbringing, almost the entire focus was on Jesus as a mystical figure. Even though there was extensive study of the scriptures, belief in his divinity and belief in Jesus as savior were much more important than what he actually said and did in his life. Emphasis was on his holiness, not on his humanness. Jesus was significant because he was the Christ, not because he was a *human* being who was exemplary in his connection with the Divine. It was of primary importance to *believe* in him, rather than to *follow his teachings*.

After having immigrated to the United States in the late sixties, as a career oriented researcher and a husband to a lovely wife and father of two toddlers, my wife and I came in touch with a movement in which primary focus was on the *teachings* of Jesus. The aspect of his *divinity* was not only de-emphasized, it was actually entirely set aside. My teachers, Emilia Rathbun and her late husband, the retired Stanford University law professor Dr. Harry Rathbun, had devoted their adult lives to communicating the teachings of Jesus, following the scholarly work of their teacher, Dr. Henry Burton Sharman.

The study of Sharman's book,[32)] "*Jesus as Teacher,*" had a profound impact on me. Sharman had written the book in the early 1900s as his doctoral thesis. He had scrutinized the Synoptics[149] for those passages that, in his scholarly assessment, withstood the test of historical authenticity and rigorously discarded everything that did not meet these standards. Many "unsubstantiated" sayings

---

[149] The books of Matthew, Mark, Luke, and – dealt with somewhat separately – John.

about Jesus are not included in *Jesus as Teacher*, including important Christian dogmas like virgin birth, bodily resurrection of Jesus, and many mystical sayings about Jesus. This clearly dwarfed my child-hood/adolescence experience of simply looking at Jesus as savior. The fact that this person, who taught in so many variations the simple message of leading a life of loving – loving God, self, and fellow human beings – was a human being like you and I, and not some supernatural figure whom to emulate was impossible, came down on me. I had no doubt that a *divine* figure would be able to profoundly convey such teachings, but to attribute them to a *human being* was truly awesome. Jesus, the example of a life lived, had become more important to me than Jesus, the one and only son of God and originator of one of the great religions.

After more than two decades of intensive studies of *Jesus as Teacher*, the modern-day mystic Ron Roth[150] brought the mystical aspects of Jesus back into my life. As a former Roman Catholic priest, Bishop Roth merged the best of his tradition with his calling as a spiritual healer. He led me back to seeing the mystical, the divine in the human person Jesus of Nazareth. Yes, Jesus is the Son of God, and, yes, he was entirely human. Yes, he is one of the most profound mystical figures ever, and yes, he taught and lived a simple and direct message that we must all adopt[151] if we want to attain the kind of consciousness we are destined to in our lifetimes. Ron Roth's devotion to the mystical does, however, not stop with Jesus. The divine can be found in other great religious teachers, in Saints – and there is a divine aspect in you and me.

Jesus was a human being – and more. You and I are human – and more!

---

[150] Ron Roth authored several books on the power of prayer and spiritual healing.[29)]

[151] This message is, as I have elaborated on elsewhere in this book, *to be a loving human being*.

I have thus come full circle in my personal belief system and am now embracing both aspects of Jesus, the wisdom teacher Jesus of Nazareth and the mystic Jesus the Christ, the Divine become flesh.

I see this development entirely consistent with the dualism principle as I have explained it in this book. Jesus is the exemplary human/divine duality. One dualism aspect is represented in his *teaching* activity, which is related to the dissemination of knowledge (albeit knowledge of spiritual matters) and concentrates mostly on the physical reality. The other dualism aspect is his *divinity*, which is related to consciousness and rooted in the spiritual reality.

Jesus is the exemplary human/divine duality. The person who sincerely studies Jesus, the teacher, will have the same chance of discovering the true nature of Jesus as the person focusing his attention on Jesus, the Mystic, the Son of God, the Savior. In both cases, the finding will likely be that the true nature is *love*.

Let us recall that a dualism, in as much as the term is used in this book, does not describe an "*either-or*" but a "*both-and*" situation. It is simply two ways of looking at the same entity. Either dualism aspect *fully* describes the entity. When studying Jesus, there is total congruence between the two. There is a seamless incorporation of both aspects in one person. The person who studies Jesus, the Teacher, will have the same chance of discovering the true nature of Jesus, the human/divine being, as the person focusing his attention on Jesus, the mystic, the Son of God, the Savior. He will have the same possibility to draw his own conclusion from his encounter with Jesus, regardless on which aspect he focuses his attention. The only thing that matters is what he actually *does* with his findings as far as his own life's direction is concerned. *If he incorporates the principle of love in his life, he will gain it. Else he will lose it.*[152] We don't all have to be alike!

[152] This is actually a free transposition of Jesus' *Great Paradox*: "*Whosoever shall seek his life shall lose it; but who loseth his life shall find it.*" It was Harry Rathbun's greatest joy to translate this Paradox into plain English, something like this[27)]: "*A person who builds walls around his psyche with the intent to protect it, will utterly lose the little bit of identity or personal worth he might be having; but the person who tears down these walls of false protection and allows his true self to emerge and come into being, will bring forth a vibrant thing. He will have gained life at its fullest." (See also the last item in Chapter 7.2).*

CHAPTER 6

Expansion of Healing

Be prepared at all times for the gifts of God and be ready always for new ones. For God is a thousand times more ready to give than we are.

Meister Eckhart

6.1 The Healing Power of the Spiritual Realm

> ***Prayer is not asking for things – not even for the best things; it is going where they are.***
>
> Gerald Heard

One of the apparent great powers of the Spirit (i.e., the gnostic reality, the spiritual reality, the divine) is its capacity of *healing*. Ron Roth,[153] a contemporary mystic and renowned spiritual healer, emphasizes that "the Spirit *intrinsically* favors health in human beings."[154] "The Spirit does not like or cause people to be sick."

The Divine *intrinsically* favors health in human beings.

We often assume that, since the Spirit is so powerful, it will do the necessary healing out of its own initiative. Ron Roth and many other spiritual healers[155] point out that this is not so. It is necessary to *invoke* the Spirit to action. Viewed from the perspec-

[153] Ron Roth, Ph.D., is the author of several books on this subject, such as *Holy Spirit, the Boundless Energy of God.*[29)]

[154] Dr. Harry Rathbun, coming from an entirely different perspective, made essentially the same statement: "The Universe is favorable for the development of life."[27)]

[155] See, for example, "*The Secret Teachings of the Espiritistas,*" by Harvey Martin[24)].

tive of the spiritual reality as I have described in this book, this would make perfect sense. Temporal and spatial omnipresence is a *circumstance*, a *possibility* that exists in the spiritual reality. It does allow spiritual energy[156] to be focused *instantly* onto a certain spot or circumstance in the physical reality. But that does not happen *automatically*. When you are looking into your back yard, you will not instantly see that one specific flower that is being visited by a bee, but you will instead *notice* it only when you direct your *attention* to that particular flower. Quite similarly, the Spirit has to direct its attention. It has to be made aware. *If asked*, the Spirit will direct its healing power to a person in need of healing. *It knows, where to go. But it must be asked*!

The Spirit knows what is in need of healing. But it must be asked! You can never over-exert the Spirit. It can never be asked too much. It will never say, "Wait a while, I don't have time right now."

Omnipresence does, however, add an interesting perspective to this topic. Have you ever asked yourself how it can be that God can be with *everybody*, or with every Easter church service all over the world *at the same time*? Isn't she way too busy with the really sick people in India to be able to attend to *my* tummy ache?

As we have said earlier, the *speed* with which the Spirit would be understood as capable of directing its attention to objects/subjects in the physical reality is essentially *infinitely* high. Consequently, even though successively and not entirely simultaneously, the Spirit would be capable of directing its attention to an essentially *infinite* number of people who ask for it *at essentially*

[156] The term "spiritual energy," as I am using it throughout this book, is equivalent to "life force energy," "subtle energy," "prana," "chi," "qi," "rei-ki."

the same time.[157] You can never over-exert the Spirit. It can never be asked too much. It will never say, "Wait a while, I don't have time right now." Even if it would take its time, perhaps even a lot of time when viewed from the perspective of the spiritual reality, it would still come through as *instantaneous* in our physical reality. God's power is infinite, ubiquitous, and instantaneous!

---

[157] As we have discussed elsewhere in this book, the normal velocities in the spiritual reality may be as fast as 10^{30} times the speed of light. Consequently, if it would take a (human) Doctor to treat one single sick person in, say 30 minutes, the number of people that could be "treated" *from within the spiritual reality* with the same degree of attention in just one physical second would be 10^{27}, or 1billion billion billion people.

6.2 Holistic Healing

I make the distinction between three classes of methods to cure diseases: (i) allopathic medicine, (ii) holistic healing, and (iii) spiritual healing. Even though there are certain areas of overlap, I see these classes as *fundamentally different*.

In our Western culture, we have come to value allopathic medicine to cure diseases so much that we consider holistic healing or even spiritual healing as outright *suspect*. This is very different from how it used to be. In the very early days of human societies, curing disease was considered beyond the realm of influence of people, and if there was any healing at all, it was experienced and accepted as a direct grace from the gods, i.e., there was only "spiritual" healing. As our societies matured, we learned about the effect of the environment on our health and how to make use of the medicinal effect of plants, as well as of certain life style changes, to cure ailments.[158] While we took some of the aspects of tending to our health in our own hands, there was still predominantly a holistic approach to healing. We knew little about our bodies or the *details* of diseases, and our approach of treating ailments involved our entire bodies.

First, there was *nothing but* spiritual healing ....

It wasn't until quite recently in human history that we became more and more sophisticated in the development of chemicals and physical and surgical tools that would very specifically

[158] A prime example would be Hildegard of Bingen,[1)] a 12th century mystic who taught and wrote about the medicinal effect of certain plants. Her recipes were re-discovered in recent years and are sold and used by many people all over the world.

treat certain ailments. This development most markedly took hold in our western culture, and most predominantly during the last century. Even as little as 50 years ago, holistic medicine was still practiced quite markedly in Germany. I grew up with *Echinacea* and other homeopathic remedies. Going to the spa to rejuvenate your entire body was commonplace. Our love affair with allopathic medicine, taking pills for every ailment, from runny nose to having a harder than "usual" time paying attention in school, got started merely about half a century ago. Since then, we have gotten so much into this routine that we are taking antidotes for the side effects of the medicines we are taking.

... but ever since then we have been engulfed in the allopathic mind-set to the extent that, for all practical purposes, nothing else seems to count. We frown upon homeopathic remedies and chiropractic treatment and are embarrassed if someone finds out we have been seeing an acupuncturist or a *Reiki Master*.

It has been only very recently that this massive trend has begun to change. Once again, we are beginning to see that the art of holistic healing has its advantages. New holistic treatment modalities are sprouting like weed – or, to use a more sinister metaphor – like antigens generated by a powerful vaccine. It would be futile to attempt to list them all. Modalities I have personally been involved with[159] include *Network Spinal Analysis (NSA) Chiro-*

[159] At the time of writing this book, I have been under *Network Spinal Analysis* care for a number of years, I have been a *Reiki* Master for 4 years and a *Karuna Reiki* Master for two years, and my wife has been administering all of the mentioned modalities for a number of years in her alternative healing arts practice.

practic®,[160] *Somato Respiratory Integration (SRI),*[10),161] *Reiki,*[26),162] *Karuna Reiki®,*[30),163] *Quantum-Touch®,*[164] *Healing Touch®,*[165] *Jin*

---

[160] *Network Spinal Analysis* (NSA) was developed by Donald Epstein, D.C., as an extension of regular chiropractic treatment for patient wellness care. The practitioner uses gentle touch at specific points along the spine to open up communication channels between the brain and specific organs to "entrain" the conditions of correct organ functioning in the patient. Several hundred chiropractors world-wide have been certified in NSA by the *Association of Network Care* (ANC), which was founded and is run by Dr. Epstein.

[161] *Somato Respiratory Integration (*SRI) is a powerful healing art developed by Donald Epstein, D.C. Described in his book "*The 12 Stages of Healing,*"[10)] SRI is a spiritually based program that combines parts of ancient principles of breathing and movements with state-of-the-art somato respiratory wellness techniques.

[162] *Rei* ("universal")-*ki* ("vital life energy") is one of the most ancient healing methods known. It was rediscovered in the early 1900s by Dr. Mikao Usui, a Japanese teacher. It vitalizes life force and balances the energies in our body. The Reiki Master is used as a channel to conduct universal life energy, which accesses the innate ability of our bio-physical body to heal ourselves on all levels. Numerous books have been published on Reiki healing; the one by William Lee Rand[26)] is one of my favorites.

[163] *Karuna Reiki®* has been developed a few years ago at the *International Center for Reiki Training* as a unifying, further empowering deeply spiritually based addition to conventional Reiki. In an attempt to avoid the wide spectrum of teaching methodologies heretofore experienced with "regular" Reiki, *Karuna Reiki* training and certification is offered only by Karuna Reiki Masters who have agreed to teach it in strict adherence to the original guidelines.

[164] *Quantum-Touch®* is a method of effective hands-on healing, developed by Richard Gordon.[15)] He states that "*by 'running energy' into the body, all healing is accelerated. The energy affects matter on a subatomic or quantum level and works its way up to the cells, nerves, and muscle tissue of the body. Eventually, the energy even affects bones to glide back into alignment. Such an adjustment tends to be longer lasting than one that came about by forced 'conventional' treatment.*" Quantum-Touch® has been successfully applied in challenges with internal organs, systemic conditions, and to alleviate pain, trauma, and inflammation due to injury or surgery. Overall, the process of healing tends to be substantially accelerated when Quantum-Touch® is administered. The principles behind Quantum-Touch® involve resonance, intention, attention, breath, and innate body intelligence. Practitioners learn to "*hold a high vibra-*

Shin Jyutsu®,[166] *Intuitive Energy Medicine®,*[167] *and Interactive Guided Imagery®*, to name just a few.

Without prejudice, it is fair to state that many of the teachers and practitioners of these holistic healing modalities tend to

---

tional energy in their hands. The recipient will eventually match the vibrations." When two energy fields vibrate at different frequencies, there is a tendency for harmonizing (*"entrainment"*), resulting in the lower frequency increasing, the upper frequency decreasing, or both meeting in-between. In this process, a person's innate body intelligence directs the healing. See also www.quantumtouch.com.

[165] *Quantum-Touch®* is a method of effective hands-on healing, developed by Richard Gordon.[15)] He states that "*by 'running energy' into the body, all healing is accelerated. The energy affects matter on a subatomic or quantum level and works its way up to the cells, nerves, and muscle tissue of the body. Eventually, the energy even affects bones to glide back into alignment. Such an adjustment tends to be longer lasting than one that came about by forced 'conventional' treatment.*" Quantum-Touch® has been successfully applied in challenges with internal organs, systemic conditions, and to alleviate pain, trauma, and inflammation due to injury or surgery. Overall, the process of healing tends to be substantially accelerated when Quantum-Touch® is administered. The principles behind Quantum-Touch® involve resonance, intention, attention, breath, and innate body intelligence. Practitioners learn to "*hold a high vibrational energy in their hands. The recipient will eventually match the vibrations.*" When two energy fields vibrate at different frequencies, there is a tendency for harmonizing (*"entrainment"*), resulting in the lower frequency increasing, the upper frequency decreasing, or both meeting in-between. In this process, a person's innate body intelligence directs the healing. See also www.quantumtouch.com.

[166] *Jin Shin Jyutsu* is an ancient hands-on healing art rediscovered in the early 1900s by Jiro Murai, a Japanese philosopher, and perfected in life-long work by Mary Burmeister. The official website www.jinshinjyutsu.com presents great detail about this modality. It states, *"The purpose of Jin Shin Jyutsu is to release the tensions that cause various physical symptoms. The body, Burmeister teaches, contains energy pathways that feed life into all cells. When one or more of these paths become blocked, the damming effect can lead to discomfort or pain. Jin Shin Jyutsu, like acupuncture and acupressure, re-harmonizes and balances the energy flows.*"

[167] *Intuitive Energy Medicine* was developed by Suzanne Louise. For a description of this healing art, see www.healingguidance.net.

emphasize *technique* over *content.* They place emphasis on perfection of the *technique* used, rather then on where the healing *power* comes from and what inner/attitudinal condition the practitioner (and the patient) must fulfill in order for the healing to be most effective.[168] In order to be allowed to administer *Network Spinal Analysis Chiropractic*, for example, one must be a licensed Doctor of Chiropractic and have passed an extensive study course program offered by the originator of the modality, Dr. Donald Epstein.[10)169] If you want to become a certified practitioner in, say, *Quantum-Touch*®,[170] you must become proficient in certain breathing techniques. In *Intuitive Energy Medicine*, one must take a week-long course and adhere to certain written texts as prerequisite for becoming a licensed practitioner. In *Jin Shin Jyutsu*, several week-long courses and demonstration of detailed knowledge of the position of a variety of "energy locks" in the human body and skill in applying hand positions on them are required. *Healing Touch*

---

[168] I want to accentuate, though, that it is the *emphasis* on technique that prompted classification under "holistic healing," rather than "spiritual healing," of these healing arts modalities. Many practitioners of these arts are (also) devout *spiritual* healers.

[169] Among the holistic healing modalities mentioned here, NSA is clearly the technically most sophisticated and most demanding art, and the requirement of a completed study program in chiropractic as a prerequisite for NSA certification seems quite appropriate. Nevertheless, it appears to be NSA founder Dr. Epstein's goal to eventually establish Network Spinal Analysis as a powerful healing modality entirely separate from chiropractic. This would eliminate the restrictive prerequisite for practitioner certification and open up the field of potential NSA specialists to talented healing arts practitioners who do not have the formal D.C. degree. While I mention NSA here as a prime example of a technique-oriented healing modality, Dr. Epstein himself, as well as many of the hundreds of NSA practitioners world-wide, are devoutly spiritually based in their motivation and attitude. It is my firm belief that it is just this unique combination of spiritual healing and extensive clinical/psychological research of Dr. Epstein and many of the licensed NSA specialists that make NSA one of the most effective alternative healing arts I have experienced.

[170] Among the modalities mentioned, *Quantum-Touch*® shines among those that offer reasonably priced courses and modest certification process requirements.

practitioners have undergone an extensive training program, in which they have learned to administer certain precise hand and finger positions for treatment of specific ailments. And so it is, in principle, with many of the numerous other alternative healing art modalities not mentioned here. As a general rule, emphasis is on *doing it correctly*, such as placing the practitioner's hand on the right spot, on applying the correct pressure, or appropriately inhaling and exhaling, or verbalizing the right words.

Emphasis on technique, rather than on the true *origin* of the healing energy passed on to the patient, is not, in and by itself, cause for criticism. It is simply the dividing line between what I understand under "holistic" and "spiritual" healing. Both categories have merit, both are powerful, both should be practiced, and patients should seek healing from both. However, problems do arise if practitioners from either category look down upon the other because, in their judgment, that they do not use the "proper" technique. The problem is particularly severe if it is a professed *spiritual healer* who gets preoccupied with technique over content.

There is also a great number of healing arts practitioners and instructors who do clearly see the connection of their specific modalities with spiritual healing and teach *to get out of the way* and *let the Spirit do the work.* For them, the modality is only a gateway for spiritual healing being set in motion.[171]

Deepak Chopra has stated, ***"Consciousness is the stuff of the universe."*** With that notion, one can no longer dismiss the validity of holistic healing. A certain degree of consciousness, the imprint or knowledge of correct functioning is available in each cell in our bodies. If the communication channels are open, there is no reason that healthy cells cannot and will not impose on un-

---

[171] Again, I mention here as an example the originator of *Network Spinal Analysis*, Dr. Donald Epstein[10)] (see also Footnote [169]). While very concerned about applying his healing art with the proper technique, he is a deeply spiritual person, and there may be few indeed who have a better grasp than he does about the true *source* of all healing that he facilitates.

healthy ones, anywhere in the body, the "consciousness" to stop acting up and shape up. Communication is the key. The cells in our bodies are not just there; they are not individual agglomerations of molecules without a context. They are not unrelated to their milieu but are in a constant communicative context with other cells. They "know" what their purpose is and what they are supposed to do. In fact, it could be argued that cancerous cells may well be those that have somehow lost their knowledge about the context they exist in. Having lost their consciousness, they are vegetating in their surroundings. Still having the capability of multiplication by cell division, they grow into something that is still of biological nature but has no more usefulness in the context they find themselves in, such as the human body.

Holistic healing is, thus, all about maintaining and/or re-establishing the consciousness about correct functioning of the cell communities throughout our body. It would seem absurd to assume that this would *best* and *only* be done with chemicals that we introduce into our bodies in the form of "medicine." Instead, it would seem highly logical to assume that our mind, the "center of our consciousness," has a significant role in the healing process. It operates in the very realm where healing has to occur – re-establishment of broken communication links.

Holistic healing is maintaining and/or re-establishing the consciousness of the cell communities throughout our body.

Many, if not most of the "modern" holistic healing modalities fit this pattern and are successful, because they address the innate capability of the body to "heal" itself[172] by fixing broken

[172] E.g., Richard Gordon, the originator of *Quantum-Touch*,[15)] refers to the person being healed (*not* the healing arts practitioner!) as the "healer." He thus clearly recognizes that a healing event obtained with *Quantum-Touch* is primar-

communication links within its parts.[173] However, identification of the healing energy as coming from an abundant, ubiquitous non-physical[174] *source*, is often not the holistic healer's primary concern and/or interest.

---

ily *not* an achievement of the practitioner (nor of the healing method itself), but solely a result of the patient's innate healing capability. (See also Footnote [164]).

[173] Of those I mentioned in the text above, *Network Spinal Analysis* (NSA) is the most fitting example of a holistic healing method. The terms *entrainment, innate intelligence, establishing communication with the brain*, belong to the key vocabulary used in NSA.

[174] I use the word "n*on*-physical" in the sense of not being subject to the physical limitations of the speed of light. The reader may be reminded that, for reasons of simplicity, I have been defining the realm of existence where process occur with super-luminous velocities as *non*-physical, while modern physics treats all existence, regardless of such limitations, as non-dualistic "physical" reality.

6.3 Spiritual Healing

> *More than 130 controlled laboratory studies show, in general, that prayer or a prayer-like state of compassion, empathy, and love can bring about healthful changes in many types of living things, from humans to bacteria. This does not mean that prayer always works but that, statistically speaking, prayer is effective.*
>
> Larry Dossey, M.D.

Spiritual healing is a method of holistic healing that does not depend on form but is entirely based in content. It has to do with intent, attention, attentiveness, focus, consciousness, awareness, the inner attitude of the healing practitioner. I will be more explicit about this later on and will, for now, deal with the issue of *receptiveness* or *appropriateness* for spiritual healing.

> **Spiritual healing does not depend on form but is entirely based in *content*, which has to do with intent, attention, attentiveness, focus, consciousness, awareness, the inner attitude of the practitioner.**

Whereby the end-result of spiritual healing – as far as the physical body is concerned[175] – is the same as that of holistic heal-

[175] This condition is added, because spiritual healing, more so than allopathic

ing, which is re-establishment of the cells' correct functioning, the way to get there is quite different. In spiritual healing, no emphasis is – or should be – placed on following a particular *form*. The only ingredient that really matters is the practitioner's *intent*. The practitioner must be practicing true love and must, without any reservation, intend the best outcome for the patient.

The basis and prerequisite for spiritual healing is unconditional love and pure motivation of the practitioner.

This may sound quite easy but is far from that. The definition of *true love* has been the subject of countless studies, and much has been written about it. True love, as I see it, implies absence of egocentric involvement. It is unidirectional (toward the patient), unconditional, and wants no reward.[176]

While the exercise of "true love" is indeed a stringent condition, sainthood is not a requirement for spiritual healing. The marvelous framework of existence which we are part of allows us to become conscious, wise, compassionate, and ultimately loving beings even within our fallible human nature.[177] Nevertheless, if we apply these standards, there is obviously a scarcity of truly "*qualified*" spiritual healers. Consequently, it is no wonder that spiritual healing lacks notoriety when compared to the other categories of healing (allopathic and holistic). Rightfully or not, the

---

and even holistic healing, also addresses the other levels of human existence (emotional/psychological, mental, and spiritual).

[176] Jesus of Nazareth, one of the greatest spiritual healers of all times, has set the ground rules of what love in this context means. In what is known as the "First" and the "Second Commandment," he states: "*Thou shalt love the Lord, thy God, with all thy soul, with all thy mind, with all thy heart, and with all thy strength*;" and, "*Thou shall love thy neighbor as thyself*."

[177] See Chapter 4.3 in this regard.

spiritual healers' motivations are often questioned. They are particularly met with skepticism when they take remuneration for their services and, more generally, don't "fit" the picture of a Mother Theresa.

One of the most effective spiritual healers of the physical body in our time is a Brazilian man commonly known in the English speaking world as "*John of God*."[178] He performs physical healings on hundreds of ill people *each day*. He does this without pay and without discrimination, to rich and poor and people of all religions – the only requirement of the people he heals is that they must be *seeking* to be healed. People do this by traveling to the small village in Brazil, where Joao de Teixeira de Faria works, and by following the instructions given to them before and after the healings. When Joao performs the healings, he does this in a trance and will typically not remember what went on during the healing sessions. He even performs surgery, when the condition of the patient warrants it. Such surgeries are performed under primitive circumstances, with simple knives, and without anesthesia.[179] His patients feel no pain, even during and following the most severe surgeries. They are, however, advised to refrain from certain strenuous activities, and they are required to maintain a strict diet following the healing treatment.

It is said that Joao's *Spirit Guides*[180] actually perform this healing work. They would use his body to administer treatments as they would commonly be done in their own reality.[181] Joao is

---

[178] Joao Teixeira de Faria, see reference [24)].

[179] This has actually been the cause of tremendous opposition by his country's medical establishment. Nevertheless, Joao's clean record of success has acquiesced this opposition and granted him tacit approval to continue with his style of healings.

[180] See Chapter 7.

[181] ... which would be what I have called the *counterpart* or *spiritual reality*.

the medium. In an exemplary way Joao passes the credit for his enormous healing power on to where it is due, not accepting any praise for himself. And it is precisely this humility that I believe assures Joao's continued effective healing work. It is precisely this clean attitude that makes him a true *spiritual* healer.

A similarly impressive spiritual healer is *Ron Roth.*[29)] As a former Roman Catholic priest, he discovered his gift and calling as spiritual healer some 20 years ago and has been teaching, healing, and inspiring ever since. Similar to John of God, Bishop Roth performs his healings under spiritual guidance and is often unaware of what is happening in detail during his spiritual healing services. He praises and credits the *source* of the healing energy, the *Holy Spirit,*[182] for the miraculous work he is performing.

Dr. Roth, more so than any spiritual healer whom – or of whom – I have known, emphasizes that spiritual healing does not only touch and heal the physical body of the patient. Quite often, he explains, what primarily needs healing is not of a physical nature but pertains to the "patient's" emotional/psychological, mental, or spiritual aspects of being. More so than in healings of physical ailments, the sacred nature of the person and of this spiritual healing process become apparent. Ron Roth performs hundreds of spiritual healings per year, often in audiences of several hundred people.

Quite often, spiritual healing addresses not only the physical but also the emotional-psychological, mental, or spiritual aspects of being of the patient.

Ron Roth also emphasizes that the gift of administering spiritual healings is open to anybody who sincerely seeks to learn

[182] … which is, as I see it, synonymous with what I have called the *counterpart* or *spiritual* or *divine reality.*

it. Having myself had the privilege of going through his multi-year ordination program,[183] I have personally witnessed hundreds of spiritual healings he has facilitated. I have seen people being wheeled into the room in wheelchairs and pushing their chairs out the room. I know people whose cancer was declared "in remission" after Ron's healing.[184] And, most importantly, I have seen many people who were healed from various sorts of serious and obvious psychological problems.

The gift of administering spiritual healing is open to anybody who sincerely seeks to learn it.

Spiritual healers, such as John of God and Ron Roth, emphasize that spiritual healing must typically be followed by a change of the particular element of the patient's life style that contributed to contracting the ailment to begin with. John is particularly strict about this and requires a 4 to 6-week specific diet after his healings, followed by strong recommendations of continued, albeit less meticulous dieting[185] and/or an appropriate physical exercise program.

Some of the commonly practiced holistic healing modalities[186] include spiritual healing as a major element, and many holistic healing practitioners are spiritual healers. This is, for example, often the case for *Reiki*.[162] At the beginning and at the end of each Reiki session, the traditional, devout Reiki Master affirms

---

[183] Details of this program can be found in www.ronroth.com.

[184] Ron Roth would joyfully respond to this by saying, "*Remission – that's the word doctors use when they really want to say "what do I know what happened? The cancer was there, but now I can't find it anymore*!"

[185] Details of what Joao Teixeira de Faria ("John of God") requires and recommends can be found in the above referenced book.[24)] The diets appear quite sensible and include certain legumes and all-organic foods.

[186] See previous chapter.

that he or she is nothing other than a *channel* for the *Universal Life Energy* ("rei-ki") to flow to the recipient and become available there for physical healing. The Reiki Master will not claim that he is a healer, but only that he is a *facilitator* for the healing process. Dr. Usui, his successor Dr. Chujiro Hayashi, in turn his designated successor Mrs. Hawayo Takata, and many of the Reiki masters in their lineage, were and are deeply spiritual persons who perform Reiki healing from the clean motivation of unselfishly serving others.

The healing or "rei-ki" energy (or the "qi" in *Qigong*) that is being activated is nothing other than what people with predominantly Christian background call "the energy of the Holy Spirit." It is also nothing other than the universal energy field that I have called "entity X" in Chapter 1[187] and subsequently re-named "*Supreme-Original Energy*."[188] This energy is abundantly available; it permeates *everything everywhere* and is universal.

I understand the healing effect of laying hands in strategic positions on the body of the recipient as a focusing mechanism of the universal energy field, not unlike a piece of iron that bends and "sucks up" magnetic field lines. The use of certain ancient healing symbols in Reiki complements the healing effect.

The *International Center for Reiki Training* has in recent years promoted an even more powerful variation of Reiki healing which they call *Karuna Reiki*® (see Footnote [163]). Having myself completed *Karuna Reiki Master* training, I found that this healing modality, even more so than "regular" Reiki, fits the definition of spiritual healing. The importance of hand positions is clearly surpassed by emphasis on attention and intention focused on the recipient. The effectiveness of the symbols used in the treatment does not depend on how well they are drawn but on the practitioner's state of being.

---

[187] See Chapters 1.3 and 1.8, and Tab. 1.2.

[188] See Chapters 1.9 and 1.10 with Fig. 1.1.

The most direct proof that spiritual healing is based on non-physical[189] energy comes from the evidence that *healing from a distance* works.[190] Over 170 scientific studies had been reported by 2002 on the effect of distant spiritual healing, i.e., under conditions where *no physical influence of the healer on the patient existed.*[191] One of the most significant of these was a double-blind study on a large number of advanced AIDS patients by Elizabeth Targ, M.D.,[34),192] reported in the peer-reviewed Western Journal of Medicine. The body of information obtained from these studies indicates that physical closeness or contact between healing practitioner and patient is all but immaterial. In most cases, the actual *form* of distant healing that was performed was left up to the practitioner. Distant healing is also practiced as a component of many alternative healing modalities, including *Quantum-Touch* and *Reiki.* It transcends the differences between these modalities and unifies them under one umbrella that one might choose to call *focused healing intention, focusing prayer, intercessory prayer*, or simply *spiritual healing.*

---

[189] Again, what is meant here with "non-physical" is "not conforming to classical physics that excludes everything that occurs at velocities greater than the speed of light and cannot be reproducibly measured with conventional instrumentation."

[190] See also Chapter 6.5 on this subject.

[191] In fact, the studies often went into great detail assuring the elimination of such influence. Often the elimination of *physical* influence was not the only consideration, and the protocols used to ascertain objectivity of the studies included measures to exclude psychological effects, as well.

[192] Neither patients nor their nurses and doctors knew details about the study, certainly not who was performing the distant spiritual healing and which patients were involved. The results indicated that those patients who received distant healing had less pain, needed less medical care, experienced fewer complications from their ailment, and had generally improved health and quality of life compared to the control group.

6.4 *Caveats* in Spiritual Healing

On the basis of the evidence, if spiritual healing were a drug, it would be on the market!

Dianne Wardell and Daniel Benor[40)]

During recent years, the practice of spiritual healing has become so prolific that I feel compelled to devote a separate chapter to the subject of discernment and *caveats* when it gets to spiritual healing. I concede that it is highly presumptuous for me to render a judgment about what may and what may not be *correct* spiritual healing. However, the subject is too important to me not to be audacious and stick my neck out. Bear with me, when I am stepping on your toes, and please remember the old saying that there is always an exception, and *the exception proves the rule.*

First, there is the question of *charging* for spiritual healing services. The perception of many recipients of such services has been that there *should* be no charge.[193] When it gets to religious/spiritual services, many people expect it *should* be free, or – more precisely stated – feel it would somehow be unethical to charge for such services.

---

[193] This perception is certainly less prevalent in the U.S. than in other countries. Having grown up in Germany, I am still sensitive to the – albeit often unspoken – notion that certain services are *expected* to be at no or very minimal charge. Such services include those of religious/spiritual, humanitarian, and sometimes even medical care nature. In fact, I grew up in a church that, for many years, had no paid staff and *entirely* relied on volunteerism. The well-known Indian born Intuitive *Mother Meera*, who has been working in Germany for many years, does her work of divine love and compassion *at no charge*. People flock to her beautiful place in a small village near Frankfurt, and about 200 men and women receive her *darshan* in 2-3 hour sessions offered four times each week.

This topic has also been of concern to many holistic and spiritual healers. While many would feel more at ease if their services could be rendered free of charge, they face the reality that they have to make a living. Often they have spent considerable resources for training and certifications in their healing art and have given up their "conventional" careers to be able to devote full-time to their new vocation, and they really have no choice but to charge for their healing services.

At the same time, our society is so entrenched in measuring value received with money spent that paying for spiritual healing services actually has a therapeutic effect. People will take a healing session, or even a course in a healing art, more seriously when they have to pay for it. Its perceived value is enhanced by the price tag.

Each practitioner will have to come to peace with this issue in his or her own way, and the answer will be different for everyone. Factors to consider are economic necessity, the cost originally incurred while acquiring the skill and certification, associated costs, such as insurance and rent for the treatment facility, and a myriad of other economic factors. It is certainly far too simplistic to judge a healer's motivation and effectiveness from looking at his or her financial practices.

Each practitioner will have to come to peace with the issue of charging for their spiritual healing services in his or her own way, and the answer will be different for each of them.

One can find the entire spectrum from rendering their healing services free of charge to charging very high fees among devout, genuine spiritual healers. For example, when *Reiki* was first opened up to practitioner training outside of Japan, no less than $10,000 was charged for Reiki Master training. Nowadays, common rates are around $500-1000, and some Reiki Masters offer "initiations" at even lower charge. In some instances, such as in the more recently developed arts of *Karuna® Reiki* (see Footnote [163]) and *Quantum-Touch®* (see Footnote [164]), instructors are actually required to abide by a certain fee structure as part of their instructor licensing agreement.

Then there is the issue of *defining* what spiritual healing actually means.[194] As *New Age* philosophies are spreading in our society, so is the use (and misuse) of the word "spiritual." Some people will call you *spiritual* simply when they find out that you are an environmentalist; for them the dolphins are spiritual beings, and even the flowers and the wind are "spiritual." In a way, you can, of course, define the word "spiritual" so that this becomes acceptable. Certainly, since *everything* is part the *Divine Oneness*, this would be true also for the dolphins and the redwood tree and the dandelion.

However, my personal definition of spirituality, as I am using the word throughout this book, is much narrower. To me, spirituality is the state of inner connection of a human being with the divine ("spiritual") reality. This is an *active, participating* relationship, the human being the participant who must do his part. If I like watching PGA golf tournaments on TV, I am not necessarily a golfer. I must actually play golf in order to be a golfer.[195]

Spirituality is the state of inner connection of a human being with the divine reality. This is an *active, participating* relationship; the person is the participant and must do his or her part.

[194] See Chapter 3.2. as a background to the term *spiritual reality*.

[195] However, I don't have to be as good at it as Tiger Woods. That's not the requirement, either.

To me, the spiritual reality is what is *primary*, with its omni-present, omni-temporal, and omniscient characteristics. A *person's* spirituality is secondary to that; the ability to connect with the divine spirituality is given to us as a *privilege*. This privilege, or grace, includes that, by virtue of our innate cognitive abilities, we can – albeit in a limited fashion – even *function* in the spiritual realm in our day-to-day physical lives.

As I have pointed out in various places in this book, our spirituality is *tied to our practice of brotherly love and compassion*. In the great scheme of things, it is then up to us to seek or not seek worldly, ego-gratifying rewards, to respond or not respond with love and compassion, in short: to be or not be spiritual. I would prefer to simply identify a spiritual healer as a person who is in this state of altruistic love.[196]

It is up to us to seek or not seek worldly, ego-gratifying rewards, to respond or not respond with love and compassion, in short: to be or not be spiritual.

Another caveat is *projections*. It is quite natural for us to project onto other people. On a person with an M.D. degree we would project that she knows something about medicine. If she authored a book in her field of specialty, we defer to her authority even more so. As "simple" healing arts practitioners, we might perhaps even feel intimidated just *talking* to her. We tend to be submissive or overly reverent to persons in high office, simply be-

---

[196] Such a person may, of course, still be a great medical professional, who is effective in his or her field of allopathic or holistic medicine. My skepticism applies specifically to persons who profess to be *spiritual* healers and don't appear to aspire to an attitude of unconditional love in the execution of their vocation.

cause we project preeminence. And we would probably defer to a religious dignitary as being particularly *spiritual*, because we would project that characteristic onto a person in that office.

Such projections may or may not be justified.[197] Especially when it comes to spiritual healing, the profession, degrees, fame, status, popularity, eloquence, or other "qualifications" of the healer may not necessarily be the most fitting indicator of his or her effectiveness as a *spiritual* healer.

Yet another caveat is *expectations*. As it is with allopathic and holistic medicine, there is no *guarantee* that spiritual healing produces the expected result. Our spiritual healing efforts, so we tend to rationalize, are *successful* when the patient is physically healed. However, in the greater scheme of consciousness and life, our human rationale is certainly inadequate to determine what would be the best healing result in a particular circumstance. As hurtful and sad as it might be, particularly when the person in physical distress is a loved one, *physical* healing may turn out *not* to be in the patient's best interest.

When the renowned physician Elizabeth Targ, the principal author of one of the most publicized and affirmative studies of the healing power of prayer,[34)] died from a brain tumor in 2002, many of her friends were confronted with a difficult-to-accept situation. The intense focusing of healing energy by countless people for her

---

[197] A number of years ago, shortly after my wife and I had immigrated to the U.S., we joined a Christian church in our new hometown. For some inexplicable reason, we were both almost immediately asked to become members of the board of deacons. From our frame of reference, this seemed like a great honor to us. Back in my fundamentalist church in Germany, one would go through great efforts to select only the most dignified members as elders of a congregation. It took but a couple of board of deacons meetings to realize that I had made an entirely false projection onto our new church. It wasn't what I was used to. I felt I was a member of the board of a social club, with preoccupations about membership drives, fund raising campaigns, and budget overruns, rather than concerns about the spiritual well-being of the people.

recuperation *appeared* to have been "unsuccessful." This *appearance* was certainly correct when measured against the *expectation* that this remarkable person, who had literally *legitimized* spiritual healing in the conventional medical society, "should" experience healing when she herself needed it.

As it turned out, we were taught a lesson – a lesson that *specific* expectations of results are incompatible with the principle of spiritual healing, and a lesson that *there is more to Life than life.*[198] In fact, as it appears to be unfolding at the time of this writing, some indications of the deeper meaning behind her seemingly untimely passing are unraveling.[199]

---

[198] I have discussed this "Great Paradox" at the end of Chapter 7.2.

[199] This observation, which I have intentionally kept vague, is based on a private communication about a year after Elizabeth Targ's passing with the renowned scientist, author, and spiritual healer Dr. Jane Katra,[35)] who was a close friend and confidante of Dr. Targ.

6.5 Spiritual Healing, Proof of Spiritual Reality?

Over 170 scientific studies have been performed and published (through 2002) on the effect of spiritual healing.[200] The body of information obtained from these studies statistically supports that spiritual healing is effective.

If there is any "proof" one would want to see for the very existence of the spiritual reality, the mere fact that spiritual healing *does exist* and has been documented over and over again seems to deliver that proof.

The fact that *spiritual healing does exist* seems to present proof that a reality exists in which consciousness reigns and which has at its avail an energy that can affect bio-physical matter in a way that appears to be independent of physical space and time. The picture painted of the "spiritual" reality with this evidence at hand is entirely congruent with my conclusion about the counterpart/divine/spiritual reality, as I have described it at the beginning of this book.

This is, in fact, the reason why I have included the chapter on expansion of healing in this book. Any person who reads about the life and work of John of God,[24)] or studies the work of Elizabeth Targ,[34)] and has a reasonably open mind would be hard-pressed not to accept at least some of the reported healings as authentic. Consequently, he would *have to* lend some validity to the existence of the spiritual reality.

---

[200] See also footnotes 191 and 192.

For me personally, however, the sequence was reversed. The hypothesis of the counterpart reality, based on an extrapolation of the wave-particle dualism, came first and awakened a keen interest in expanding it to "explore" the existence of a spiritual reality. Spiritual healing became a logical consequence of that train of thought.

The dualism work actually dates back to 1979 and occurred, to a large extent, in the form of automatic writing.[201] It took almost two more decades[202] until I came to the point of mental concession that spiritual healing irrefutably works.

It is my deep wish that this book may lead people with similar conditioning to a reduction of this excessively long mental gestation period toward acceptance of the "nobler hypothesis."[203]

---

[201] Most of Chapter 1 of this book was conceived during a three-day period in the winter of 1979.

[202] Until October, 1997, at a seminar on spiritual healing with Ron Roth.[29)]

[203] See Chapter 7.1, question on *faith*, quote by Harry Rathbun.[27)]

CHAPTER 7

Conversations

In this chapter I am responding to some pertinent questions that have come up in the context of the material presented in the preceding chapters. The answers, even though often given in quite a different, more "colloquial" language, give examples of the conclusions I have drawn from this material.

While I used restrained language in the preceding chapters, in an attempt to purposefully avoid conclusions that might be conceived as going "too far," so I would not turn off the scientifically trained reader with statements that are coming from the heart rather from the mind, I am now pulling out all stops in the conversations recorded in this chapter. Here, my answers are coming straight from the heart. Even though most details are congruent with the mental framework presented in the preceding chapters, I make no attempt to restrain myself when congruency becomes a bit "sketchy."

7.1 On the Physical Reality/Spiritual Reality Dualism

Q: *You use the analogy of the particle-wave dualism to deduce characteristics of the spiritual reality. Would you re-cap the main characteristics of the spiritual reality?*

A: First, the physical reality and the spiritual reality are dualism aspects of the Ultimate reality, which I have called "Supreme-Original Reality," similar to mass and waves being dualism aspects of something higher, which is "energy" in that case. All occurrences in the spiritual reality happen at super-fast velocities, as many as 30 orders of magnitude greater than the speed of light. Consequently, for our human frame of reference, all things are eternal in the spiritual reality, including the life span of the human's counterpart in that reality, the soul. What "energy" is for us in the physical reality would be "thought" in the spiritual reality. "Knowledge" here would be equivalent to "consciousness" there.

Q: *Is there a difference between the spiritual reality and what is called "heaven" in the sacred scriptures?*

A: I think they are one and the same. The word "heaven" is ancient and reflects the best understanding of the concept of a divine reality available at the time. Obviously, there was no dualism law in physics to draw from. People didn't know what the speed of light is. In fact, there was no physics. One knew essentially nothing about the universe and how it works. Since the sky, with the moon, the sun, and all the stars, was entirely out of human reach, it was something very mysterious, and it is easily understandable that it became the place where God (or the Gods) resided. It became the location for the spiritual reality. Now that we know more, we find the spiritual reality to be all around us, part of us, actually literally within us, and we can make all the assumptions about it that I have presented in this book. But what I have really shown is that the essence of our primitive assumptions about heaven, which we have held for three millennia, is still valid today. The essence is that a spiritual reality, or a heaven, does exist and is the realm where a consciousness vastly superior to human consciousness reigns. So, when I verbalize the first affirmation of the "Our Father," I acknowledge this mysterious divine reality and my own "significant insignificance" within it.

> **The *essence* of our primitive assumptions about heaven, which we have held for three millennia, is still valid today.**

Q: *Tell more about this "significant insignificance" that we have in the spiritual reality.*

A: The significance is that with our reflective mind, we actually function in, or penetrate into, the divine reality. We are divine beings. And with that attribute, we can, to a limited degree, influence occurrences in the spiritual reality. Our consciousness is part of that realm. The *in*significance is that the divine reality is so much greater, so much more powerful, so much more significant than we humans are that it would be utterly arrogant to assume that our individual roles in that realm are overly important. I have used the example of the importance of individual cells in our body as a metaphor. We are for the divine reality what our cells are for our bodies – not insignificant, but not overly important either.

Q: *Are there, as you see it, discrete entities in the spiritual reality?*

A: As I stated, the human being and the soul form a dualism, each being "confined" to their respective realms. Since we have societies of human beings in the physical reality, it would be a logical extension to assume that there are societies of spiritual beings in the spiritual reality. They may be composed of souls, spirits, angels. These may, in fact, just be synonymous words. There may even be intricate hierarchies of spirit beings. This view is corroborated by many authors who reported about messages perceived or received from "the other side," including the prominent psychic Sylvia Browne.[4)]

Q: *So, might an Archangel be a more elevated being than, say, a "regular" soul?*

A: For a number of decades of my life I have thought that *spirits*, *angels*, *archangels*, are nothing but un-contemporary, if not naïve, religious constructs. This thinking has changed. They may be very real, perhaps – or even likely – a lot more real than you and I are as human beings. This is the long preface to a simple "yes" to your question. I consider it plausible that the societies in the spiritual reality are headed by a few especially highly evolved spirits which some of our forefathers, for whatever reason, have identified as archangels. Remember, there is no life/death cycle in the spiritual reality, as we know it. The archangels of the Old Testament times would still exist today, and would continue to exist in infinity.

Q: *How sure are you about all this?*

A: Of course, this is speculation. Nobody knows for sure. In fact, the very circumstance that we cannot be sure about those things is part of the ingenious Great Divine Master Plan. This is because, if we could really ***know***, without a shadow of a doubt, the Plan would not work. It is based on free choice. If we ***know***, then there is no choice. Then there is just stupidity if we don't do what we know is right. In all of this, the emphasis is on the word "know," "knowing beyond any doubt," just like knowing that 2+2=4.

Q: *What about re-incarnation? Does the concept of reincarnation fit into your construct of the spiritual reality?*

A: First of all, I find your choice of words unfitting. To me it has become much more than a "construct." It has become a solid faith....

Q: *...What is faith?*

A: My esteemed spiritual teacher, the late Dr. Harry Rathbun, would say this: "Faith is choice of the nobler alternative!" For me, given the old-fashioned opinion that nothing exists outside of our physical reality and the modern suggestion that a spiritual reality is likely to exist, the latter is the nobler alternative.

Now then, back to the question of reincarnation. Since I consider it very likely that the spiritual reality is vastly superior to "our" physical reality, I must conclude that it is conceivable that incarnation in a physical life is an option, or a possibility that a being in the spiritual reality has. It might choose the option in order to gain further experience. The intended outcome of this experience would be an increase in consciousness.

Q: *Why wouldn't it simply be to experience fun, or misery?*

A: When we go back to our contemplation of the spiritual reality as the dualism counterpart to the physical reality, we recall that there is an intrinsic dualism counterpart to the energy → entropy process, which we identified as the thought → consciousness process. So I think that it would not be just fun for the sake of fun, or misery for the sake of misery, but it would be fun or misery ***for the sake of becoming more conscious***. I cannot conceive of this entire *Divine Plan* unless it has consciousness as a major purpose. This does not mean that we should feel guilty if life brings us fun and happiness! Not at all! But what it does mean is that, whatever life does offer us, there is an element in it that our soul wants us to learn from the experience. What we make of the fun, or the misery, or whatever life presents us, will make all the difference!

Q: *Getting back to the discussion on the spiritual reality, are there other realities, in addition to the spiritual reality?*

A: **O**ur physical reality exists in a framework of very few basic natural physical constants, such as the speed of light, electron charge, proton mass,[204] electron mass, Planck's energy constant, gravitational constant, which allow every physical thing and process to occur in essentially "empty" space. Don't forget: the atoms that make up our physical world consist of a nucleus that has minute size but contains essentially all the mass and, thus, energy of the atom, and an electron charge whose primary function it is to keep the nucleus of neighboring atoms at their proper distance. That distance is relatively ***huge***. Compare, if you will, the carbon-monoxide molecule (CO) with Earth and Mars. If the planets are represented by the nuclei of the oxygen and the carbon atoms, the space in-between Mars and Earth compares to the space between of the two nuclei in the CO molecule. Everything in-between is "empty" space. It is quite conceivable that another "physical" reality that is based on a different set of natural constants could entirely co-exist with our physical reality in that same "empty space." In fact, modern physics considers the notion that just one reality would exist, with just one set of natural constants, as limiting and thus unrealistic and ***suggests*** the existence of parallel realities. They would entirely co-exist and not interfere with each other, similar to how radio waves from many different stations can interference-free co-exist in the same space.

---

[204] This is only an approximation. It assumes that the proton and neutron masses are the same (which is true within 0.1%) and does not consider the known sub-elementary building blocks, leptons (including neutrinos), quarks, and mediators.

Compare the carbon-monoxide molecule with Earth and Mars. If the planets are the nuclei of the oxygen and the carbon atoms, the space in-between Mars and Earth compares to the space between of the two nuclei in the CO molecule. Everything in-between is empty space.

Q: *Would this mean that there are many spiritual realities?*

A: Each of these would be a *physical* reality. And each of these physical realities could then have a spiritual reality as dualism counterpart.... But enough said about this. Our one physical reality, and our "attached" spiritual reality, are big enough for us to deal with! After all, our physical reality alone includes the entire universe, with billions of stars in our galaxy alone, and billions of galaxies out there ...

7.2 On Reincarnation ("Study Trip" Analogy)

> ***We're not human beings that have occasional spiritual experiences – it's the other way around: we're spiritual beings that have occasional human experiences.***
>
> Deepak Chopra

Q: *Talk more about reincarnation, how would this work?*

A: The analogy I am going to describe may seem a bit far-fetched, but I think it does have some merit. I would venture to speculate that for a spirit entity to incarnate into a human being for a physical lifetime might be comparable to a person going on a study trip, say, to a foreign country. First of all, the trip would be for a finite length of time. "Return" to the traveler's "regular" home life would be part of the plan, would be expected. In fact, the return date would likely more or less be pre-arranged. However, there would always be the possibility that something important comes up that changes that date: a change of circumstances, an illness ... Furthermore, there would be a specific ***purpose*** for the trip, such as learning certain new aspects concerning the traveler's profession....

For a spirit entity to incarnate into a human being for a physical lifetime might be comparable to a person going on a study trip.

Q: *...and if that learning did not meet the expectations, another trip might become necessary to make up for what was missed the first time around?*

A: Precisely. A follow-up trip might also be considered when the first trip went exceedingly well. Why not consider repeating a successful experience, perhaps with a different itinerary, to further broaden the experience? Why wouldn't you, after a number of trips undertaken for *learning* purposes, consider going back on travel to ***teach***, for a change? So you have multiple reincarnations and the emergence of true spiritual teachers built into the system.

Q: *Might Jesus have been a reincarnated master teacher?*

A: It would certainly make sense to see it that way! The Buddha, Mohammed, many of the Christian Saints, Yogananda, and modern-day mystics and Gurus might fall in that category. In a way, this would cast a new light on our authority problem! It would explain that there are truly inspired, enlightened human beings living side by side with us ordinary people. For me, this thought has been somewhat of a revelation. While I never had a problem recognizing and accepting the real great spiritual teachers, such as the ones I just mentioned, I have had my hard times with some of the modern-day religious authority figures. To bow my head in reverence toward a priest, a bishop – or even the Pope himself – was not entirely congenial to me. The realization that these people, even though humans like you and me, might be incarnations of exceptionally evolved spirit beings – or extremely high-ranking individuals of the Spirit Societies – definitely tumbled my mental block toward accepting these individuals as higher evolved beings, certainly worthy of my deference.

Q: *What other analogies would you draw between the study trip and (re)incarnation?*

A: You can go on and on spinning with this analogy. The trip widens your horizon – increases your consciousness. You make use of what you learned on the trip when you are back home – our consciousness has a beneficial effect on what is going on in the spiritual reality. When your mission is completed and you are ready to go back home, people whom you have gotten to know during your study excursion may be sad that you are leaving them, probably feeling that they will never see you again – which, of course, would put a new spin on our perception when a loved one dies, or "passes on," as one quite fittingly would say in certain spiritual circles. Certain activities back home, such as the direction of R&D at your place of work, may literally *depend* on conferences that bring many people working in your field together for a mutually beneficial scientific exchange. Many would thus plan their trip to occur at the same time, so they can meet with certain other people. In analogy, one would have to assume that the learning a soul yields from a human lifetime could be quite important for the soul's spiritual society.

Q: *Would there be predetermination? Would the learning experience be pre-determined?*

A: You would make your airline and hotel reservations before you leave, wouldn't you? You would prepay your tuition or conference fees. You might even pre-plan your agenda in great detail, such as the events you want to participate in and the people with whom you want to meet on your trip. These agenda items are then *somewhat* predetermined – but you have the option and can exercise your free will to drop your agenda at any point on your trip. At that moment, you are throwing all pre-determination over board, and you are thus proving that there really is no predetermination when it comes to human beings. You may also encounter unforeseen circumstances and simply ***decide*** to change your plans. You may also get into an accident and mess up your itinerary altogether. Pre-***disposition***, certainly; but not pre-determination! It is a wondrous system which we are part of! I sometimes wonder what the world is really like that I will be coming back "home" to when my field trip into the physical realm is over.

Q: *You make it sound like as if life in the Spiritual Realm is the "really real" life, and life in the physical realm the more tenuous, more "sketchy" one of the two.*

A: I think you are right on. I wouldn't go quite as far as scholars of the eastern illusionist philosophies put it when they flatly state that our physical life is an illusion. But I do believe that it may well be more appropriate to say, "The human is an incarnation of the Spirit," thus emphasizing that the Spirit is primary, rather than, "A human has a spirit or soul," which would convey the notion that the human may be primary. Of the dualism "physical reality/spiritual reality," both exist and have their place, but the latter is the more powerful reality, because it works with an additional dimension, one that puts it beyond time and space limitations.

Q: *Did Jesus know about all this?*

A: There is no doubt in my mind that Jesus was an incarnation of one of the – if not *the* – highest "ranking" beings in the Spirit world. He lived a life and said things that clearly indicate that he was conscious of all of these things, while living in a place and at a time where essentially nobody around him had even a sliver of such a highly developed awareness. What sense would you make out of this saying of Jesus: "***If a person seeks to save his life, he will lose it. But if a person loses his life, he will find it.***"

Q: *What did Jesus mean when he said this?*

A: This is a somewhat free "wording" of what is known as the "Great Paradox." In different Bible texts you will find slightly different wordings, all being an attempt of interpretation by the translators of what Jesus *might* have meant with his saying. A widely accepted variation is replacement of the last word, "it," with "eternal life." Others elaborate on the first half of the second sentence by adding something like "for my sake" or "for God's sake," which would render that sentence to *"... But if a person loses his life for my sake, he will gain eternal life."* Expanding on the interpretation of my great spiritual teacher, the late Stanford University law professor and exemplary scholar of the "Teachings of Jesus," Dr. Harry Rathbun,[27)] I would interpret the Great Paradox this way:

> ***"If a person builds up walls around his psyche in an attempt to protect himself, if he tries to attain fulfillment by amassing worldly goods and/or fame, if he sets his entire focus on physical wellness, he will, in the end, utterly fail. But if he takes down the protective walls he has built around his psyche and gives his true nature of a compassionate and loving being a chance to emerge, if he detaches from his preoccupations with worldly possessions, fame, or physical well-being, so his function as a compassionate and loving person will not be contaminated with nonessential preoccupations, he will bring forth a vibrant conscious being that will live in eternity."***

Yes, Jesus did know about the spiritual realm. He knew that he was an incarnation of the Spirit of God, and he knew that you and I are, too. He knew that his mission was to teach. He knew that there is no greater teaching than that which makes us realize that practicing compassion and love is the highest function of a human being, and it is what our lives are all about.

7.3 On Conditions for Growth in Consciousness

Q: *In Chapter 4.3 you state that there is a good reason why a person's soul cannot directly change the person's physical life. You say that, if direct interference of the soul in its human counterpart's physical life were possible, this might jeopardize the significance of the decisions that the person makes and, thus, undermine the degree of consciousness the person could achieve in his lifetime. You state that the decisions, if they are to add up in an advancement of the person's consciousness, must be taken in freedom of thought. How could it happen that the soul interferes with freedom of thought?*

A: By ***authoritatively*** telling us what to do. If our soul would tell us, without leaving any doubt, what the correct action in a certain situation would be, there would be no ***real*** decision necessary; we would just do it, simply because we would ***know*** it's right, and we would be stupid if we wouldn't do it. There is then no more choice. There is no more alternative. It's like having a glass of plain water and a glass of water laced with poison in front of you. If you ***knew*** which one contains the poison, would you choose to drink that one? Not if you are sane! So, knowing precisely what your choices entail, you have no more freedom. The outcome is determined. It is true, however, that a course of events coming about with the help of your inner voice, or your soul telling you unmistakably what to do, might help get a certain good cause accomplished – but that particular activity would do little to advance your ***consciousness***. Consciousness arises from an altruistically motivated ***decision***, a real decision for the best result between genuine alternatives.

Q: *Are you inferring that our freedom of thought might be at stake if we listen to our inner voice?*

A: Not at all! Our inner voices, our hunches, our intuition are probably the most valuable indicators we have on our life journey. However, if they were ***more*** than that, if they were ***real data***, information that you can ***unmistakably prove***, they would indeed not be too helpful to ***create consciousness***. They might be useful to solve problems, at times important problems we might be facing, but they would do little in terms of advancing our consciousness. Solving physical problems and developing consciousness are two different propositions. It's very simple: it does not take a rocket scientist to figure out that one plus one equals two. That's just a fact, and you would be stupid – plain stupid, not just "unconscious" – if you would not ***always*** obey that law. Acting as if one plus one would equal three might get you in deep trouble. You wouldn't pay for three loafs of bread if you put only two in your shopping basket, would you? Paying for two rather than three may have the very worthy benefit of keeping you financially afloat, but it does nothing to enhance your consciousness. But when it gets to what it might mean to respond with one little act of compassion to one person in need, we all too often shrug our shoulder and pass up this perfect opportunity to add a bit of consciousness to our soul.

Q: *So you are making a distinction between knowledge – and the good that can come from it – and consciousness and its distinct benefits?*

A: Very much so! There is a lot of good that can come from knowledge. Let us be totally clear about that. The world would still be in the dark ages, had it not been for the enormous amount of knowledge we have collectively achieved. But let us also be clear that our technological advances are based on our mental achievements – not on our consciousness. That is something different. We did not land man on the moon because of our consciousness, we did it on account of knowledge. Granted, there were many leaps in consciousness that tagged along with this great technological achievement, but that is a different story. I recall Rusty Schweickart's often cited realization[205] when he circled around the moon, "***On that small spot, that little blue and white thing, is everything that means anything to you – all of history and music and poetry and art and death and birth and love, tears, joy, games, all of it on that little spot out there that you can cover up with your thumb***." That was not knowledge, that was a leap in consciousness!

Q: *Are you then saying that, when it gets to growth in consciousness, it may not be the most desirable approach to consult a psychic, or to strife to attain that ability yourself, so you would know the answers to your questions?*

A: I think that you would certainly stand at risk of not advancing as much in consciousness as you would if you were living your life in such a way that your important decisions are coming straight from the depth of your heart, so-to-speak "unadulterated" with what people with psychic abilities suggest you should do.

Q: *Does this imply that it is part of the Grand Divine Plan that we should not know and thereby not deliberately seek information about our past lives and/or our future?*

A: In general, I prefer to act from an inner decision for the good, rather than from a notion derived from a reading of my past or future that a particular response will be beneficial for me. While such a reading might satisfy my curiosity or even increase my ***knowledge*** or worldly recognition, it might ultimately deprive me from growing in consciousness.

---

[205] Russel Schweickart was the Lunar Module pilot for the Apollo 9 earth-orbital flight in March 1969.

I'll give you a personal example that might make this clear. For many years, I had felt intrigued by the ability of certain people to know about their past lives. Some would talk about who they were and what they did, almost as if they were talking about what happened to them a few years ago. I thought I had absolutely no such ability, and I often felt inferior because of it. Then, one night I had a flash-back – not a regular dream, but without any doubt a "real" flash-back – of a situation from a past life. It was as clear as if I was re-living it. I saw myself in the town square of a medieval town, a place I had never been to in my (present) life, in a situation I had never even closely been in before; a situation that had never before come up in a dream. The scene was horrible. Plain horrible. I was about to be fatally stabbed in a political uprising. I was begging for clemency, but to no avail.

I was then able to consciously pull myself out of that past life situation and immediately knew that something profound had been revealed to me, something that would give me insights into why I am the way I am. But I also knew instantly that this knowledge was not ***essential*** for me, and I felt that, if it continued, it could actually have a negative effect on me. My intuition, or my inner voice, told me what to do next. I remembered what I had learned from spiritual teachers about such occurrences and firmly asked the Holy Spirit to engulf me with white light and surround me with a protective energy field that would disable any future attempt of any spirit entity, whatever it might be, to intrude into my present life with inappropriate information from past lives. My inner peace was immediately restored, and the incidence remained "forgotten" until – quite appropriately – it resurfaced this very moment as an example in this conversation.

Q: *Is then, in your opinion, clairvoyance a questionable human ability?*

A: Clairvoyance is a great gift and certainly not negative, but clairvoyance that is not spiritually based should be approached with great discernment. The recipient of a clairvoyant reading has the important task of deciding what to do with the information divulged to him. There is a wise rule that certain things cannot be undone. We cannot really pretend we never heard what we heard. I was once quite innocently offered the opportunity of a reading into my future. Since this was a new experience for me, I somewhat naively went for it. I was told that in August, in three months, there would be a significant change in my relationship with my wife. That thought kept bothering me. August came, and August passed. Luckily, the change did not come to pass – or I missed it. But for three months, the thought of this impending worrisome change had preoccupied me.

With regard to a person's ability to be clairvoyant, the challenge will usually be discernment in the use of this ability with other people.[206] It can be used as a gift or as a weapon, and/or as a means to make an honest or not so honest living.

Q: *Aren't you closing yourself off by denying the all-out benefit of clairvoyance? Aren't you falling in the trap of going against progress?*

A: I don't think so. I am not suggesting that it is ***wrong*** to get a psychic reading! All I am saying is that it may be that this activity does not help advancing my ***consciousness***. We do many things that are not designed to advance our consciousness. In fact, the vast majority of things we do have little or nothing to do with consciousness. They may still be necessary and important, but they

---

[206] I recently had a discussion with a person on this subject who is an extremely brilliant, widely known author, innovator, and holistic healer. After having known him for several years, he confided to me that he has the gift of unusually detailed and accurate clairvoyance. Because of his concerns for the person whose life story and problems may have been revealed to him in great detail through clairvoyance, he would typically communicate only the bare essentials of that knowledge. He would do this only in an intentionally tentative way, such that the person would not even become aware of his psychic abilities.

are not an issue involving the advancement of our consciousness. It may be perfectly OK to consult a psychic for any issue giving us greater ***knowledge***, giving us factual information – as long as we are clear it's not consciousness but ***facts*** or ***knowledge*** were are after. A person I have known quite well is a highly gifted clairvoyant psychic who gives readings on people's character. You can phone him up and ask him to give you a reading on another person. All he needs to know is the name and, perhaps, a location, and he can zero in on that person and give you certain information about him that you might want to get clarity about. I once chose to get his reading on a person with whom I was about to enter into an important business relationship. The result was extremely helpful – and correct. It helped me advance a ***particular physical situation*** – but it did not do much in advancing my consciousness!

Q: *The Divine Plan is then ultimately not about knowledge but about consciousness – it is not about doing good, but about deciding to do good?*

A: It's about deciding to do good ***and then doing it***! But you are right, the element of decision makes all the difference, when your growth in consciousness is at stake! If, when faced with alternatives, you are able to ***rationally*** determine the outcome that is most advantageous for you – even if this "rational" fact helping your determination comes from an "irrational" psychic reading – there is no more ***real*** decision to make, and you are bypassing your opportunity to grow in consciousness. You are then disenfranchising yourself of an opportunity for growth. But when you allow the mystery to stay in its place, then you can make a conscious decision and grow in consciousness.

I really think that this is precisely the issue that we "Westerners" have to come to grips with in these times. It has been way too long since we have confused knowledge with consciousness. We have revered the great scientists, engineers, people who possess great *knowledge*. We have revered them because of what they *know*, not for who they *are*. We have been at awe about the brilliance of a mind, not about the *wisdom* of the person behind the mind. Yes, there is a difference between knowledge and wisdom!

Q: *But is it not true that knowledge and wisdom go hand in hand?*

A: Quite frequently that is the case, but unfortunately, in the big picture in western civilization it appears to be the exception, not the rule. There are quite a few examples of people with brilliant minds who also possess great wisdom. Clearly, Albert Einstein was a wise man. Jonas Salk, the brilliant inventor of the polio vaccine, was a man of great wisdom. Stephen Hawking, one of the world's foremost astronomers and theoretical physicists, who came closer than anybody to a theory describing "everything," openly admitted his intuition of a great mystery underlying everything. Robert Oppenheimer, the father of the atom bomb, became deeply disturbed about his brainchild. And this list goes on and on.

But again, the exception proves the rule, so the saying goes. Sadly enough, the rule is still that probably the vast majority of brilliant engineers, scientists, computer specialists, even physicians are satisfied with the advancement of *knowledge* they contribute through their work, regardless of the societal value it generates. Most of us are still in the paradigm of being more impressed with a person's possessions then how much wisdom he or she possesses – all these are, sadly enough, still the norm, not the exception.

Overall, in the big picture, we must begin to *value consciousness above knowledge.*

7.4 About the Role of Mystery in our Lives

Q: *Say more about the role of mystery in our lives. What is its place? What is our appropriate response to mystery?*

A: We must then first clarify what you mean with "appropriate" – appropriate for what? If you are designing a vehicle to shuttle a group of people to Mars, you would probably be well advised not to lean too heavily on the Mystery but rather rely on ***knowledge***. If someone asks you for directions to the airport, you probably won't respond, "let the Spirit guide you."

I would say that Mystery is of great importance for ***growth in consciousness***. You will do a great service to your soul, if you realize that there ***is*** mystery. We must realize that we are but insignificant beings in a vastly superior, mystical context. The context is mystical because it is far beyond our capacity of mental understanding. We would fare well if we recognized the strength and power of the Great Mystery that surrounds us and permeates us. We would be wise if we associated with this mystery the utmost degree of goodwill toward us, the utmost benevolent directionality and will. But we would be foolish to tamper with this mystery for the sole reason of satisfying our curiosity, or otherwise attempting to profit from it by (ab)using it to get ahead of someone else. The Mystery is part of us. Life itself is a mystery. Mystery sustains us. Therefore, becoming increasingly aware of the Mystery is important. But this activity requires reverence and humility, the two key ingredients for our response to the great Mystery.

Q: *How can I become more aware of the Great Mystery?*

A: Seek teachers who strike you as living in reverence and humility. Don't act as if you know it all. Forget the notion that only what you can see, hear, touch, smell, calculate, construct is real, and nothing else exists. That notion is not even contemporary any more. It is out-dated! Let the awe get a hold of you.

Q: *The awe for what?*

A: The awe for your life, for beauty, for how things are. The awe that you are able to see me! I am composed of essentially nothing but empty space, and so are you! The mere fact we can see each other, in spite of the emptiness, is awesome! And in colors! That's awesome.

You are wondrously made! Think about it. The billions of cells that make up your body aren't just sitting there, they know precisely what to do! When you cut yourself, they know how to rebuild themselves and heal the wound, and after a few days or weeks you won't even see anymore where you had the cut. Your cells have knowledge – that's awesome! The millions and millions of new cells that your body is replacing every day in its continual regeneration process remember their exact pattern and their precise function – that's awesome! Look at the moon. What keeps it from colliding with the earth, or from disappearing into space? Gravity is awesome! Look at the sun. Its energy fuels all life in the solar system. Energy is awesome! Don't ever think for a moment that, because we know that $E=mc^2$ we know what energy is and where it ultimately comes from. Don't ever assume that, because we can calculate the effect of gravity, we know what gravity ***is***!

Yes, we are wondrously made, the world, the universe is wondrously made! Look at your body. That knee joint that you use thousands of times every day is one of the most ingenious "mechanical devices" ever made. No human has ever invented a mechanical device that would operate a hundred years without ever wearing out. Look at your eye. Can you think of making one in a factory? And your brain? It's not just a supercomputer that is designed to speedily and expediently process information. It can process information ***and*** feelings, sensation, intuition, and with that it is a tool to produce consciousness! That's worth a bit of awe!

Q: *So the Mystery is that we cannot really explain the* ***origin*** *of the basic driving forces of the Universe?*

A: More than that. The mystery is the enormous intelligence that must have been there, at the very beginning of it all. This intelligence did not start with the Big Bang, it was already there eons before the Big Bang occurred – if one can even express this in time frames like "eons," because we know that there was no time before time began … The Intelligence that I stand in awe before had already ***embedded all potentiality*** for the awesome evolution that would start with that big beginning point of time, space, and matter. That's the real Mystery! Our intellectual understanding of the entire physical reality, from the Big Bang to today – even if we were able to understand all intricacies, would still have to stand in awe before the Mystery that seeded the Universe with all its potentiality. ***Original Thought, Direction, Power*** – that's the Mystery.

Q: *What, do you think, is the main reason why so relatively few people truly believe that there is a "Power Greater"?*

A: I would first state that this question refers much more to us Westerners than to those who grew up, for example, in Asia. Now then, it's our education – or what we think our education tells us. It's a misapplication and a misunderstanding of knowledge. Forgive me when I commit yet another transgression in stereo-typing, in order to bring the point across: of all major Western professions, I have found that engineers are the most likely to doubt the existence of Mystery. I think that this has to do with the method of their training, which focuses on ***calculability***. Engineers learn mathematical and engineering formulas and how to put them to work. The more, the better. For them, what cannot be calculated cannot exist. Interestingly, physics, once considered the epitome of the engineering sciences, has made a full turn-around, and many of the most renowned physicists are now not only admitting that Mystery exists but actually ***postulate*** it.

We must change our Western approach to education. We must not only focus on data but include ***values***. We must not only go for knowledge but build a kinship for expansion of consciousness into our educational system.

7.5 On Practicing Spiritual Healing

Q: *What is Spiritual Healing?*

A: Let me start with saying what it is not. It is not hocus-pocus, it is not an irrational attempt to heal what conventional medicine cannot heal, it is not primitive, and it is not something for me but not for you, or vice versa. Spiritual healing is an activity of tapping into the universal energy field for the benefit of healing mind, body, and spirit.

Q: *What is it based on?*

A: Again, let me start what it is not based on. It is not based on a *human* construct. It is not the result of something we humans have invented or imagined and, if you don't believe in it, it won't apply to you. It's not the product of wish-thinking. It is based on something that is and always has been and that has absolutely nothing to do with you or me: the "Rei-ki," the universal life energy, the (energy of the) "Holy Spirit," the "Prana," the "Chi," the "Qi," or whatever you might want to call it. It is there to tap into. For you and me. It ***is***.

Q: *In previous chapters you have presented a framework that explains some characteristics of the spiritual reality. Does it also explain how spiritual healing works?*

A: I would even prefer to say that it ***suggests*** such an explanation. The framework I've presented is intended for those of us who need explanations, who are caught in the dilemma of doubting the existence of anything that we cannot see, hear, touch, smell, or calculate. It suggests that the spiritual reality is not only real, but that it is even more real than our physical reality with which we are so intimately familiar.

Q: *That's a hard statement to swallow!*

Look! What corresponds to "energy" in the physical reality – a driving force we can put to good use in our physical lives – would be "conscious thought" in the spiritual reality. It is subtle energy that can be ***directed***. Consciously directed! That's what spiritual healing is all about: ***consciously directing energy***. The energy that is being directed is not just physical energy. It is a higher-level energy. It is the kind of energy that sustains our lives and many other things that exist but that we can hardly fathom. This spiritual energy can heal – most appropriately – not only our physical bodies, but also the mental, emotional-psychological, and spiritual aspects of our being.

Q: *How can healing energy be directed? Who does the directing?*

A: By conscious thought. I use these words deliberately, because it is an action in the spiritual reality, even if it originates with a human being. There is human thought, and there is divine thought. As I have pointed out in previous chapters of the book, the realm of conscious thought is where the divine and the human meet. We as humans can direct conscious thought into the spiritual/divine realm to affect a healing action by divine thought-energy. It is as if we would be making the divine thought presence ***aware*** of a certain situation that needs to get "their" attention. If your child has been stung by a yellow jacket, and you are aware of what happened, you can do what you know you must do to ease the pain. But if you didn't notice what happened and, consequently, are unaware of your child's misery, you are not helping. So it is with the Holy Spirit, it needs to (and wants to) be made aware of a need and will then direct the healing energy toward where It thinks it's needed.

Q: *Does the recipient of the healing energy have to be in tune with this energy, does he have to believe in spirituality, in order for the healing energy to work?*

A: I think this is not necessary. There are numerous studies and experiences that suggest that a person's inner state of being is not a determining factor for receiving healing. I believe, for example, that the prayer of a mother for her child in need will be heard regardless of the child's state of spiritual awareness. And the famous study by Elizabeth Targ[207], who established the benefit of positive thought in a scientific double-blind study on a large sample of AIDS patients, clearly shows that the recipient's belief structure is not of primary significance for spiritual healing to work. I am sure it helps, but it is not essential.

Q: *What about the practitioner, does he have to be a spiritual or religious person?*

A: Yes and No. Certainly he does not have to be a ***religious*** person. In Chapter 6.3, I have suggested that the requirement for the practitioner is to be in a state of unconditional love and pure motivation. Another expression may be ***benevolent intent***. I would qualify a person who practices benevolent intent toward another person as a spiritual healer. It is immaterial if he or she goes to church, or to the temple or synagogue, or if he tithes. It also doesn't matter if he expresses his benevolent intent in the form of Qigong, Reiki, Quantum Touch, Therapeutic Touch, etc. etc., or just simply "conventional" prayer. ***The intent is what matters***. If there is one thing that Jesus of Nazareth was clear about, it is that!

---

[207] Elizabeth Targ, M.D., in her convincing peer-reviewed study in the Western Journal of Medicine;[34)] see also Footnote [192].

Q: *When intent is so important, why even bother with adhering to a certain form or technique? You have categorized Reiki as a predominantly spiritual healing modality. However, you are using certain hand positions and symbols in Reiki healing. Aren't these merely form? How can they, compared to the requirement of benevolent intent, even be significant?*

A: I think both are additive. The use of certain hand positions, combined with an impeccable inner attitude of the practitioner, can have the effect of concentrating healing energy fields into discrete areas of the body. I would compare this to the effect a piece of iron has on a magnetic field. With the mere physical shape of the iron you can influence the field so much that it can become a very powerful instrument. For example, the key elements of an electron microscope are its electromagnetic lenses, which are sets of precision-machined iron "pole pieces" with rotational symmetry. Even minute deviations from rotational symmetry of the pole pieces can distort the imaging capacity so much that the instrument is essentially worthless as a microscope.

This is how I see the role of the practitioner's hands in Reiki healing (see Footnote [162]) or some of the other healing modalities that use the positioning of hands on the recipient's body, such as Healing Touch (Footnote [Error! Bookmark not defined.]), Quantum-Touch (Footnote [164]), Jin Shin Jyutsu (Footnote [166]), or even Network Spinal Analysis (Footnote [160]), to name just a few. They shape the energy field lines in such a way that the divine healing energy[208] will be focused on the area of the body in need of healing. What the magnetic lens and the imaging electron beam are for the electron microscope, the positioned hands and the healing energy field are for the recipient of hands-on healing. If the symmetry is perfect – i.e., if the inner attitude is immaculate – the result will be excellent. Already relatively slight imperfection drastically weakens the outcome.

[208] The expression "divine healing energy" is, as mentioned in various places of this book, synonymous with what I have called "entity X" in Chapter 1 and subsequently re-named "Supreme-Original Energy."

So, to complete the answer to your question, I would say, in a simplistic way, that the purpose of the benevolent intent is to get the flow of healing energy going, for which you are the conduit, and the purpose of positioning your hands is to direct ***where*** the flow is going, or ***focus*** it.

Q: *Can you feel the energy flowing from you to the recipient?*

A: Some healers can feel it more than others. Rosalyn Bruyere,[5)] a well-known teacher, minister, and healer, who has taught extensively on healing of the chakra energies, compares the flow of subtle energies through her arms and legs to the flow of rushing water. Richard Gordon,[15)] the originator of Quantum-Touch,[209] also describes a very definite sensation when subtle energy is "running." Practitioners of subtle energy healing arts like Reiki, Healing Touch, Therapeutic Touch, Qigong, Quantum-Touch, and alike usually feel the energy much less succinctly. They typically report tingling in their hands, sometimes a sensation of electric charge coming from their fingertips, and sometimes no particular sensation at all.

Q: *Can one say that "the more, the better," i.e., the more of a definite sensation of energy running you have, the better the healing will be?*

A: I personally think that, in the context of spiritual healing, there may well be little such correlation. Clearly, if you can really strongly ***feel*** the energy running through your arms and hands, as Rosalyn Bruyere reports, this will tend to give you a lot of ***confidence*** as a healer. It will help eliminate the doubts that we all have, one way or another, that this "surreal" thing is really "real." Those of us who have less of a physical sensation when they do healing have to take their confidence from less "obvious" indicators. But other than that, I think that there is really very little dif-

---

[209] See Footnote [164].

ference between the healing performed by a healer who senses the healing energy strongly rushing through her body and one who meets the prerequisites of genuine love, as I explained earlier, but does not have such a clear physical healing sensation. How else would you explain the well-documented beneficial effect of distant healing, which is usually the result of concentrated mental focusing on the patient, rather than the laying on of hands?

Q: *What are the "less obvious" circumstances you are talking about?*

A: ***Personal experience*** that healing works. This is really the crux of the matter. Many of us have experienced absolutely marvelous healing stories, either as recipient or "practitioner" of healing arts. But, no matter how convincing the healing was at the time, how often do we somehow rationalize the story down to a random occurrence, or forget it altogether? I could come up with literally dozens of such personal healing "success" stories that all ended up in this category. In fact, as I have described elsewhere in this book in detail, I believe that it is part of the ***Great Divine Plan*** that we have to come to a point of genuine, compassionate love ***without knowing in detail how this all works***. So, the best advice I can give you in this regard is to "count your marbles" when they are dealt. Be open to see what is happening to you or your patient or a loved one. Don't allow your rational mind to immediately water it down. And then, since this watering down is all but too natural, take up some kind of a discipline that will remind you of your marbles at times of doubt. For me, a good discipline has been to write down wondrous occurrences when they are still fresh in my mind, and to re-read them when the doubts try to sneak in.

Q: *Is there a way to quantify, or scientifically prove, the existence of subtle energy?*

A: The fact that *distant* spiritual healing works is primary proof – because it takes *energy* to affect a bio-physical change. In other words, no matter how you look at it, physical healing requires energetic input. That input may be "regular" or "subtle" energy. Obviously, if you are talking about distant healing, you have by conventional definition not employed physical energy, because what is coming from your mind does not, by conventional "wisdom," have the nature of physical energy. The only conclusion is then that "subtle" energy was involved.

But I'll give you a simple different example. Have you ever felt the energy emerging out of your fingers? If you haven't, try this: open your left hand (the non-dominant hand; reverse the sides if you are left-handed) and point the index, middle, and ring fingers of your right hand toward to the palm of your left hand, without touching. Keeping a clear distance of, say, ¼ inch, move your right hand fingers up and down in front of the palm of your left hand and feel the energy received in your left hand. It is, as if you can ***feel*** where the longest of these three fingers happens to be. You can feel the subtle energy emerging from it!

There are many other ways to prove subtle energy. Let me list just a few:

- Subtle energy fields of living organisms can be photographed with Kirlian photography. The method was pioneered in 1939 by the Russian scientist couple S.D. Kirlian and V. Kirlian.[20)] Dr. William A. Tiller, Professor Emeritus of Stanford University, has extensively published about this technique.[38)] Kirlian photography has recently been substantially advanced,[21)] and Gas Discharge Visualization (GDV) instruments are now commercially available for clinical applications.[210] The GDV technique can provide aura images that are computer generated with the combined

[210] The GDV technique has been fully accepted by medical authorities for clinical applications in Russia.

bio-field information gathered from the tips of all ten fingers of a person.[211]

- Some people directly *feel* the energy emanating from a patient by gently sweeping their hands over the patient's body at a distance of several inches. The energy is particularly strong above an organ that is not properly functioning. I myself experienced this first-hand with surprising clarity shortly after a serious accident my son had suffered. The MRI had indicated internal injuries, including a ruptured spleen and a ruptured liver. The emanation of subtle energy above his chest from the areas corresponding to the respective locations of the injured organs was quite apparent. A family friend and highly intuitive NSA[212] chiropractor, who had also hurried to the scene, described these energy fields as so strong that his hands were literally ***pulled*** toward the location of the ruptured organs. They were "crying for healing energy" that was literally ***flowing*** from his hands to the patient. After a few minutes, the energy flow notably decreased. It was as if the main healing action had already been completed at that time. (We later learned that the doctors, unaware of the chiropractor's work, were quite "pleased" that these internal injuries, normally considered very serious, had healed so quickly).

---

[211] An aura image generated with the GDV technique is not the same as an aura actually *seen* by a person who has that gift. But I understand it is quite close and possibly the best possibility yet for those who do not have that gift to tap into the *benefit* derivable from seeing auras. This benefit is generally perceived as being able to pinpoint problems, or better problem tendencies, *before* they become physically manifest in the person.[5), 21)]

[212] See Chapter 6.2 and Footnote [160].

CHAPTER 8.

Concepts and Definition of Certain Often Used Terms

In this chapter, I am addressing, in alphabetical order, a few of the key concepts used in this book. I define and look at these concepts as I understand them, viewed from the perspective of the spiritual/divine reality being the duality counterpart to the physical reality. In doing so, I am aware that my definition and understanding differs at times quite markedly from how the term is used by others.

The listing is not intended to be in any way complete. I am focusing only on those terms that I consider particular related to this new perspective. Words in emphasized italics are treated separately.

Angel

The theorem of a spiritual reality developed from the dualism principle and of the development of consciousness being the purpose of the Divine Plan, as discussed in this book, would allow the possibility of existence of discrete, conscious *entities* in the ***spiritual reality***.[213] Such entities would have to be, at least from the frame of reference of a human being in the physical reality, ***omniscient*** and ***omnipresent***. This is due to the characteristic of the ***spiritual reality*** that all processes occur at super-fast velocities. For the frame of reference from *within* the ***spiritual reality***, such entities would probably be no different than humans in the physical reality. I.e., they would be separate from each other – even though interconnected, perhaps even exist in intricate societies, and they would have a purpose of existence. It is probably neither inappropriate nor naive to extrapolate from our societies to societies in the ***spiritual reality***. It is possible that no real difference – other than perhaps in hierarchy – exists between angels and ***soul***s in this context. To speculate in any way on the *appearance* or *form* of these entities is, in my judgment, difficult, if not inappropriate. If they were able to manifest/show themselves to beings in the physical reality, which ap-

---

[213] This statement is a good example of how I want to intentionally deal with topics in this chapter: rather than even *mentioning* the overwhelming multitude of records and reports about angels throughout history, I am responding narrowly only from the framework of the spiritual/divine reality as the duality counterpart to the physical reality.

parently happens occasionally, they would probably choose an appearance/form that would make it easier to be recognized by us doubt-ridden humans. Given the vastly superior resources available in the ***spiritual reality***, compared to the physical reality, it should be expected that it would be easy for angels to appear to humans in any form they wish.

Archangel

I would understand archangels as hierarchically superior to ***angels*** or ***souls***, similar to the hierarchical differences that exist among people in the physical reality.

Channeling

The occurrence of entities from the ***spiritual reality*** speaking through human media ("channels"), usually with the specific intent of imparting certain messages they are compelled to get across to us humans at large, is heavily documented. It is surprising to what extent the general description provided in these documents about the realm within which these entities exist agrees with the counterpart/***spiritual reality*** as I have described it in this book, derived from an extrapolation of the dualism principle in physics.

Much of the essence of this concept (primarily Chapter 1 of this book) originally came to me in the form of ***intuitive*** writing, which is one form of channeling.

Consciousness[214] *Knowing in context* (from Lat. *con-scire*, to *know together*, i.e., in context). It suggests that it is an outcome, a result of a process involving the mind, which works in the form of generation, projection, and analysis of "thought." (Consciousness is thus not just *awareness*, nor is *being conscious* simply the opposite of *being in a physical or physiological state of unconsciousness*). The *thought* → *consciousness* process addresses the creative, intelligent aspect of ***Supreme-Original Energy***. Consciousness spreads with infinite speed omnidirectionally and penetrates everything in the universe. It is independent of the person from whom it might have originated and "takes up a life of its own." Consciousness is an element of the ***spiritual reality*** and is the ***dualism*** counterpart to ***entropy***.

Counterpart Reality Same as ***spiritual reality***. I have used this term in the first, more science oriented chapters of the book, where a more "neutral" expression seemed appropriate.

Distant Viewing The ability to view distant places, anywhere in the world, without time or space limitation, has been researched by various groups and individuals and appears to be well established. Research into this phenomenon has been funded over extended periods of time by the U.S. Government through grants to

[214] My use of the word *consciousness* in this book is essentially identical to Joel Goldsmith's[14)] use of the term *spiritual consciousness*.

reputable institutions, such as Stanford University, and results have been used by intelligence agencies in the solution of complicated cases.[215]

Dualism Synonymous with *duality*, a phenomenon known from physics, where it has a "*both and …*" character (the wave – particle dualism in physics states that light/energy can be *fully* described as wave or as particle), rather than an "*either or …*" connotation, such as in *opposites*. In philosophy, the term *dualism* is often used synonymous to *opposite*; in this book I am maintaining a clear distinction between these terms.[216]

Entrainment (see also ***Vibrations***). This term is very appropriately often used in the holistic healing modality "Network Spinal Analysis Chiropractic"[217] where it describes a harmonizing effect that can be achieved when the communication channels between the brain and the organs are optimized. The efficacy of an *entrained wave* appears to be depending on its amplitude, not its frequency.

---

[215] A well-known principal investigator of such studies is Russel Targ.[35)]

[216] The *physical reality* and *spiritual reality* are not extremes that have nothing in common, but they are both aspects of the same, higher-level reality (the *Supreme-Original Reality*).

[217] Developed by Dr. Donald Epstein[10)] and practiced world-wide by hundreds of chiropractors licensed by him in this holistic healing modality. See Footnote [160].

Entropy — A form of energy; known from thermodynamics as *the lowest-grade energy possible*; energy that is – albeit still *energy* – not further usable for *any* physical processes.[218] It is dead, entirely useless.

Ghosts — This would be compatible with an understanding that it may, in certain relatively rare instances (perhaps as rare as manifestations of entities from the ***spiritual reality*** occur to people in the physical reality – see "***angel***"), happen that the transition of the soul of a diseased person to the ***spiritual reality*** has not been completed. That "ghost" entity would be erring, perhaps for some finite time, between these two realities, not really belonging to either, having lost the physical body and thus being "invisible" in the physical reality, and not yet having fully reassumed the characteristics and identity of an entity in the ***spiritual reality***.

Gnostic[219] Reality — Synonymous with ***counterpart reality*** and ***spiritual reality***.

---

[218] Entropy is often also defined as *chaos*. I am not using that definition in this book.

[219] The word *gnostic* is derived from the Greek "gnosis" (knowledge). The connotation between the spiritual reality and a realm in which boundless, timeless knowledge reigns was emphasized by Teilhard de Chardin.[36)] This contrasts with the common understanding of "*gnosticism*," a religious orientation dating back to the 3rd century, in which *knowledge* is the key requirement of the *follower*.

Guardian Angel To me stories like the ones I presented in Chapter 5.2 are "proof" that this is not an empty term. I submit that we all have guardian angels. They are part of the entities that make up the societies in the ***spiritual reality***. While we are unaware of them, they are probably utmost keen to communicate with us and keep us from endangering ourselves.

Holy Spirit Synonymous with ***Supreme-Original Energy***; the driving consciousness and energy of the ***Supreme-Original Reality***. It contains energy and thought as dualism aspects. It is everywhere, penetrates everything, fuels and sustains life.

Intuition The mechanism by which our ***spirit guides*** communicate to us. Large portions of this book originally came to me in the form of intuitive writing.[220] (See also ***channeling***).

Omnipresence The concept that the entire ***spiritual reality*** is, with regard to our physical reality, omnipresent follows directly from the conclusion

[220] I did not experience this as *intuitive writing* in the very moment. This is more a conclusion I have drawn in the ensuing years. I realized that I had done nothing but writing. The words and concepts "flew" onto my notebook with an ease that I had never before experienced. There was almost no way of stopping the flux. However, I did not experience that words came to me that were foreign to my vocabulary or my general field of understanding, as has been noted by many authors who have published intuitively written texts. (To me, the most astonishing intuitively written text is the "*Course in Miracles,*"[13] which is one of the most inspiring books I have ever attempted to study).

that the ***spiritual reality*** is essentially not subject to our space-time limitations.

Omniscience

Knowing everything (from Lat. *omnis* and *scire*). Consciousness is not subject to time and space limitations. It does not disappear. It is, thus, *additive* to that which is already there. Due to the super-fast reaction speeds prevailing in the ***spiritual reality***, all consciousness that has ever been attained, anywhere in the universe and beyond, is instantaneously present and available for processing. This would make (entities residing in) the ***spiritual reality*** omniscient with regard to us living in the physical reality.

Physical Reality

The universe, everything that is subject to the laws of physics (up to "Einsteinian" physics, not including phenomena that are outside of time and space). The physical reality encompasses everything that can be seen, touched, smelled, heard, measured.

Prayer

Similar to "***Spiritual Healing***," the scientific "validity" and effectiveness of prayer is one of the major conclusions of this book. However, it is important to be on common ground with what we mean when we talk about prayer. Prayer is an attitude, not an enactment of a certain ritual at a certain time and place. It is an attitude of openness to the ***spiritual reality***, or the divine realm. It is an inner proclamation that I am part of a divine system from which I am open to receive guidance, as I am aware that it is much more powerful, ***omniscient*** and ***omnipresent*** than I am. Prayer is connection with the divine. Prayer is being in a state of loving. Prayer is the most essential ingredient for the divine consciousness development plan.[221]

Raising Vibrational Frequencies (see ***Vibrations***)

Reincarnation

The theorem of a ***spiritual reality*** developed from the dualism principle, as discussed in this book, is neutral to the concept of reincarnation. However, the person-soul dualism is a major part of the book, and it can be inferred (yet not postulated) that incarnation of the soul in several lifetimes is part of the overall divine *consciousness development plan*. I personally find the concept of reincarnation of the soul in multiple lifetimes intriguing and convincingly demonstrated.

Soul

The soul is the human person's dualism counterpart in the ***spiritual reality***. It is a person's essence; that which will remain when the physical body dies. The soul wants to grow in consciousness during a person's physical lifetime.

---

[221] It feels incomplete to deal with the topic of prayer, which is one of the highest functions of a human being, and which has been the subject of countless books and studies, in just a few lines in this context.

Spirit Guide

In as much as entities can be assumed to exist in the ***spiritual reality*** (see "***Angel***"), it would be logical to conclude that certain entities in the ***spiritual reality*** would care for certain fellow entities that have chosen to ***(re-)incarnate***, more than for others, probably not dissimilar to parents caring for their children. Those could be a person's *spirit guides*. They would be all around their earthly protégé, but they would be unable to *directly* communicate with him/her, as they cannot avail themselves to the laws of our physical reality. (See "***Intuition***"). Also, and quite importantly, such direct communication would be in disagreement with the divine consciousness development plan that places highest values on consciousness attained in *free will/decision* of a human being. If a person were to receive unequivocal, convincing notions from their spirit guide(s) to do or not do certain actions, such action would be, in a way, less significant, because it would be the result of acting on *knowledge* rather than *acting on conscience*.

Spiritual Healing

The validity of spiritual healing is one of the major inferences[222] of the ***spiritual reality*** theorem developed from the dualism princi-

[222] I used the word "inferences," because it can be *inferred*, but not *postulated* from the characteristics of the spiritual reality. The inference stems from the enormous *energy* that is attributable to the spiritual reality and extends seamless into the physical reality (see Chapter 1), whereby it can be assumed ("inferred") that by mind/consciousness action of a person some of this energy can be directed into a human being for purposes of physical healing.

ple and one of the major conclusions of this book. However, it is important to be on common ground with the *definition* of the concept. In this context, the word "spiritual" describes an essential attitude of the individual administering the healing. His/her motivation must be compassion and love. The healing energy is taken entirely from the ***Supreme-Original Reality***, and the healer is, thus, the ***Supreme-Original Energy*** (***Holy Spirit***) itself, not the person performing the act of healing.[223]

Spirituality

A person's state of inner connection with the ***spiritual reality***; not the same as *religion*.

Spiritual Reality

This word is, as I understand it, synonymous with "gnostic reality," "divine reality," "divine plane," or "heaven." It is the dualism counterpart to the ***physical reality***. The existence of the spiritual reality, and its characteristics, are the center stage of this book.

All occurrences in the spiritual reality occur at super-fast velocities, as many as 30 orders of magnitude greater than the speed of light. Consequently, for our human frame of reference, all things are eternal in the spiritual reality, including the life span of a person's counterpart in that reality, the ***soul***. There may be societies of souls, spirits, ***angels***, with intricate hierarchies. I consider it very

[223] See Chapter 6.3 and Richard Gordon,[15)] originator of *Quantum-Touch* (see Footnote [164]).

likely that the spiritual reality is vastly superior to "our" physical reality. What *energy* is for us in the physical reality would be *thought* in the spiritual reality. *Knowledge* here would be equivalent to ***consciousness*** there. It is conceivable that incarnation (see ***reincarnation***) in a physical life is an option, a possibility for a ***soul*** to gain further experience, comparatively not dissimilar to what an excursion to another planet might be for a human being. The intended outcome of this experience would be an increase in consciousness. I would consider it likely that there is an enormous amount of intelligence in the spiritual reality.

Supreme-Original

The *S.-O. Reality* is the all-encompassing Oneness that contains the entirety of the ***physical reality*** and the ***spiritual reality*** as dualism elements. From it originates everything that ever existed or will ever exist.

S.-O. Energy is the driving force in the S.-O. Reality. It is the dualism parent of physical energy and thought (energy).

Vibrations

The words "***vibrations***" and "***vibrational frequencies***" have very specific meanings in science, and these are sometimes incompatible with how they are used in the context of holistic/spiritual healing and/or New Age philosophy/spirituality. Often, the word "frequency" is used where "amplitude" would be the correct term, culminating in an indiscriminate inference that "*raising vibrational frequencies*" would be desirable.

Everything vibrates, and characteristic vibrational frequencies can in principle be ascribed to everything that exists, from the subatomic particles to living cells to entire organs. Whereby it is true that a cell (or group of cells) that is "ill" (i.e., damaged, malfunctioning) has different characteristic vibrational frequencies than a healthy one, that "wrong" set of frequencies can be lower *or higher* than that of the healthy cell, entirely dependent on the specific status of the cell. It is, therefore, incomplete to *indiscriminately* aspire to *raising* of vibrational frequencies, because there may be just as many situations where *lowering* of the frequency may be what is beneficial. *Frequency adjustment, or harmonizing*[224] would be more appropriate terms to use. See also: ***Entrainment***.

A major characteristic of the ***spiritual reality*** is that *everything* in it occurs at velocities many orders of magnitude greater than the speed of light. Consequently, the characteristic vibrational frequencies in that reality must be assumed to be also *many orders of magnitude greater* than those in the physical reality. Given this comparison, it would be

---

[224] The term "harmonizing" would appropriately imply that each oscillator (i.e., vibrating entity) has an influence on its surroundings, which influence is to come into a state of "harmony" with its surroundings (i.e., mostly the cells directly surrounding it).

congenial to talk about "*raised awareness*" or "*raised levels of consciousness*." Either of these phrases substitutes exceedingly well for the terms "raised vibrational frequencies" or "raised vibrations" in most New Age/Philosophy contexts.

The *frequency* of a wave is determined by structural circumstances, i.e., it increases or decreases with a *structural* change, such as a mutation of a cell. The *amplitude* is indicative of *strength*; it increases or decreases with power. Often, especially if ***entrainment*** of a "healthy vibration" is desired, the goal would be to increase the *amplitude* of the wave. Breathing techniques, such as used in Network Chiropractic,[225] Quantum-Touch healing,[226] Rebirthing,[227] or similar holistic healing modalities are most likely beneficial because of an increased *amplitude,* rather than a changed *frequency* of the healer's energy field.

---

[225] See footnote [160].

[226] See Footnote [164].

[227] Rebirthing is a healing technique based on breathing work. The official website www.thecosmicbreath.com gives detailed information about this art.

REFERENCES[228]

1) H. von Bingen (Commentary by Matthew Fox), *Illuminations of Hildegard von Bingen*, Bear & Co., 1985.

2) Niels Bohr, *On the Constitution of Atoms and Molecules*, Phil. Mag. **6**, 26 (1913).

3) Louis-Victor de Broglie, *Ph.D. Thesis*, Henri Poincare Institute, Paris, 1924.

4) Sylvia Browne, for example in *The Other Side and Back*, Signet/Penguin, 2000.

5) Rosalyn L. Bruyere, in *Wheels of Light, a Study of the Chakras*, Bon Productions, Sierra Madre, CA, 1989.

6) J.B. Carlton and William A. Tiller, *Index of Refraction Measurements for a Superluminal Radiation*, in Proceed. 10th Ann. Med. Symp. on "A Holistic Approach to the Etiology and Therapy of the Disease Process," Phoenix, 1977

7) Charles R. Darwin, *On the Origin of Species by Means of Natural Selection*, 1859.

8) Michael Drosin, *Bible Code II – the Countdown*, Viking, Penguin Group, 2002.

9) Freeman J. Dyson, in *Infinite in All Directions*, Harper & Row, Publ., New York, 1988.

---

228 Rather than presenting an exhaustive, scholarly listing of references, I have chosen to give only a small selection of mostly those that I have found to be particularly interesting and/or poignant reading material. I have, however, not included the very extensive catalog of books that *infringe* on the topics discussed in this book and might be "recommended reading" but were not actually used as references.

10) Donald Epstein, *The 12 Stages of Healing*, Amber-Allen Publishing, 1994.

11) Albert Einstein, in Physikal. Zeitschr. 18 (1917).

12) Stapp d'Espagnat, et al., American Journal of Physics **50**, 811 - 816 (1982).

13) Foundation for Inner Peace, *A Course in Miracles.*

14) Joel Goldsmith, *The Infinite Way,* De Vorss Publications, 1992.

15) Richard Gordon, *Quantum-Touch: The Power to Heal*, North Atlantic Books, 1999.

16) John Gribbins, *Space: Our Final Frontier,* BBC Consumer Publ., 2001.

17) Stephen W. Hawking, in *A Brief History of Time*, Bantam Books, New York, 1988.

18) Klaus Heinemann, *Consciousness or Entropy?* Eloret Press, 1991.

19) Nick Herbert, e.g., in *Faster Than Light: Superluminal Loopholes in Physics*, Dutton, New York, 1988.

20) S.D. Kirlian and V. Kirlian, *Photography and Visual Observations by Means of High Frequency Currents*, J. Sci. and Appl. Photography **6**, 397, (1961).

21) K. Korotkov, *Human Energy Field: Study with GDV Bioelectrography*, Backbone Publ. Co., N.Y., 2002.

22) Friedrich Lenz, *Vorlesungen über Elektronenoptik*, Tübingen, 1966/67.

23) Michael Linton, *The Mozart Effect*, in First Things **91**, 13 (1999).

24) Harvey Martin, *The Secret Teachings of the Espiritistas*, Metamind Publ., 1998.

25) Joan Ocean, *Dolphin Connection*, Hawaii, 1989.

26) William Lee Rand, *Reiki for a New Millennium*, Vision Publications, 1998.

27) Harry Rathbun, in *Creative Initiative, a Guide to Fulfill-ment*, Creative Initiative Foundation, Palo Alto, California, 1976.

28) Sanaya Roman, in *Living with Joy*, H.J. Kramer, Inc.

29) Ron Roth, in *Holy Spirit, The Boundless Energy of God*, Hayhouse, 2000; and several other titles.

30) Laurelle Shanti Gaia, *The Book on Karuna Reiki*, Infinite Light Healing Center, Inc., 2001.

31) Henry B. Sharman, *Records of the Life of Jesus*, Sequoia Seminar Foundation, Palo Alto, California.

32) Henry B. Sharman, *Jesus as Teacher*, Sequoia Seminar Foundation, 1935.

33) Jess Stern, *The Impersonal Life – the Little book in which Elvis Found the Light*, De Vorss Publications, 2001.

34) Elizabeth Targ et al., *CPMC Distant Healing Study: a ran-domized double-blind study of prayer and distant healing for Advanced AIDS*, Western J. of Medicine, Dec. 1998.

35) Russel Targ and Jane Katra, *Miracles of the Mind*, New World Library, 1998.

36) Pierre Teilhard de Chardin, *The Phenomenon of Man*, Harper & Row, Publishers, 1978.

37) William A. Tiller, e.g., in *Phoenix: New Direction in the Study of Man*, Vol.1, p.28 (1977).

38) William A. Tiller, *On the Evolution of Electrodermal Diagnostic Instruments*, J. Adv. Medicine, **1, 1**, 41 (1988).

39) William A. Tiller, *Science and Human Transformation -- Subtle Energies, Intentionality and Consciousness*, Pavior Publishing, Walnut Creek, CA, 1997.

40) Dianne Wardell and Daniel Benor, 13th ISSSEEM Conf., Boulder, 2003.

41) Don Yeomans, J.P.L. 1998 (in Time, 3/23/98).

42) Gary Zukav, in *The Dancing Wu Li Masters*, Bantam Books, 1979.